ERCOFTAC Series

Volume 28

ERCOFTAC (European Research Community on Flow, Turbulence and Combustion) was founded as an international association with scientific objectives in 1988. ERCOFTAC strongly promotes joint efforts of European research institutes and industries that are active in the field of flow, turbulence and combustion, in order to enhance the exchange of technical and scientific information on fundamental and applied research and design. Each year, ERCOFTAC organizes several meetings in the form of workshops, conferences and summer schools, where ERCOFTAC members and other researchers meet and exchange information. The *ERCOFTAC* series publishes the proceedings of ERCOFTAC meetings, which cover all aspects of fluid mechanics. The series comprises proceedings of conferences and workshops, and of textbooks presenting the material taught at summer schools. The series covers the entire domain of fluid mechanics, which includes physical modelling, computational fluid dynamics including grid generation and turbulence modelling, measuring techniques, flow visualization as applied to industrial flows, aerodynamics, combustion, geophysical and environmental flows, hydraulics, multi-phase flows, non-Newtonian flows, astrophysical flows, laminar, turbulent and transitional flows.

Indexed by SCOPUS, Google Scholar and SpringerLink.

More information about this series at http://www.springer.com/series/5934

Matteo Pini · Carlo De Servi ·
Andrea Spinelli · Francesca di Mare ·
Alberto Guardone
Editors

Proceedings of the 3rd International Seminar on Non-Ideal Compressible Fluid Dynamics for Propulsion and Power

NICFD 2020

Springer

Editors
Matteo Pini
Propulsion and Power
Delft University of Technology
Delft, Zuid-Holland, The Netherlands

Carlo De Servi
TES group
VITO
Mol, Belgium

Andrea Spinelli
Dipartimento di Energia
Politecnico di Milano
Milan, Italy

Francesca di Mare
Institute of Energy Technology
Ruhr-Universität Bochum
Bochum, Nordrhein-Westfalen, Germany

Alberto Guardone
Dipartimento di Scienze
e Tecnologie Aerospaziali
Politecnico di Milano
Milan, Italy

ISSN 1382-4309 ISSN 2215-1826 (electronic)
ERCOFTAC Series
ISBN 978-3-030-69305-3 ISBN 978-3-030-69306-0 (eBook)
https://doi.org/10.1007/978-3-030-69306-0

This Springer imprint is published by the registered company Springer Nature Switzerland AG
The registered company address is: Gewerbestrasse 11, 6330 Cham, Switzerland

Preface

The continuous increase of worldwide energy consumption and the need to drastically abating greenhouse emissions demand for a significant increase of the renewable share in the global mix of primary energy sources, as well as the development and deployment of novel technologies to enhance the efficiency of transportation systems and industrial processes.

Thermal engines based on thermodynamic cycles where the energy conversion process exploits unconventional fluids, namely other than steam, air, and flue gases, will play a key role in this scenario. The use of unconventional substances, however, entails that the flow behavior within the components of such engines can depart from that of an ideal gas as the underlying physical process occurs partly or entirely in the non-ideal compressible flow regime. Such regime is typical of dense vapors, flows in supercritical states or close to the saturation curve and the critical point, and two-phase vapor-liquid flows.

Examples of thermal engines where non-ideal flows typically occur are those based on the concept of the organic Rankine cycle (ORC) and the supercritical CO_2 cycle (sCO_2) like ORC turbogenerators, heat pumps, and sCO_2 power systems. Turbomachinery and heat exchangers are components of these systems whose performance is affected by the occurrence of non-ideal flow effects. Other engineering applications operating with non-ideal flows are fluid machinery for the oil and gas industry, fuel injectors for high-pressure combustors, rocket engines, compressors for refrigeration systems, and flow devices for cryogenic plants.

Given that the non-ideal compressible flow regime is characterized by strong nonlinear variations of fluid property, the accurate performance assessment and the optimal design of internal flow devices require accurate modeling of the thermodynamic and transport properties of the working fluid, as well as specialized numerical methods and tools based on highly predictive physical models. The development of models and methods tailored to non-ideal compressible fluid dynamics (NICFD) calls for a combination of theoretical, numerical, and experimental activities that can only be pursued by synergistic research efforts. The biannual NICFD conference has been established to provide evidence of the latest progress and discoveries made by the community and to share the derived knowledge, with emphasis on propulsion and power technologies. Key themes of the conference are: experiments; fundamentals; numerical methods; optimization and uncertainty quantification (UQ); critical and supercritical flows; turbulence and mixing; multi-component fluid flows; applications in ORC power systems; applications in supercritical sCO_2 power systems; cryogenic flows; condensing flows in nozzles; cavitating flows; super/trans-critical fluids in space propulsion. A Special Interest Group on NICFD (SIG-49) was established in 2019 within ERCOFTAC, the European Research Community On Flow, Turbulence And Combustion.

This volume contains the papers presented at NICFD 2020: 3rd International Seminar on Non-Ideal Compressible-Fluid Dynamics for Propulsion & Power held during October 29–30, 2020, at Delft University of Technology, The Netherlands. The conference website is: NICFD2020.

The collection of proceedings is composed of 18 papers reporting on cutting-edge research activities. The papers are divided in five thematic areas: Fundamentals, Supercritical Combustion, Numerical Methods, Rigs and Components, Experiments and Measurement Techniques. The covered topics encompass theoretical modeling of canonical NICFD flows, high-fidelity numerical simulation of reacting and non-reacting flows, turbulence and loss modeling for turbomachinery, modeling of two-phase and multi-component flows, experiments in turbine flows, and advanced optimization techniques.

We would like to thank all the participants to NICFD 2020 and the members of the scientific committee for providing guidance during the organization of the conference and for managing the revision of the papers. We are also grateful to all reviewers for their invaluable support to attain the high scientific quality of the contributions collected here. A special acknowledgment goes to Federica Ascione and Francesco Tosto, PhD candidates, who greatly helped with the organization of the event.

Delft, October 2020; Dr. Matteo Pini, Delft University of Technology, Chairman; Dr. Andrea Spinelli, Politecnico di Milano, Vice-Chairman; Dr. Carlo De Servi, VITO, Vice-Chairman; Prof. Francesca Di Mare, Ruhr Universität Bochum, Review Chairman; Prof. Alberto Guardone, Politecnico di Milano, Founding Chairman.

Contents

Numerical Methods

Rigs and Components

Experiments and Measurement Techniques

Fundamentals

Investigating the Validity of the Fundamental Derivative in the Equilibrium and Non-equilibrium Two-Phase Expansion of MM

Martin T. White(✉)

City, University of London, Northampton Square, London EC1V 0HB, UK
martin.white@city.ac.uk

Abstract. The aim of the work presented in this paper is to explore the validity and possible extent to which non-ideal compressible-fluid dynamic (NICFD) effects may have a role in the expansion of a two-phase mixture of the siloxane MM by expanding the concept of the fundamental derivative to two-phase flows. For the assumption of a homogeneous two-phase mixture under the two limiting cases of thermal equilibrium and frozen thermal states, expressions for the speed of sound are defined. Using these definitions, the fundamental derivative is evaluated within the saturation dome of MM under the assumption of thermal equilibrium. Following this, the variation in the fundamental derivative under thermal equilibrium and frozen thermal states are compared for expansion processes from different reservoir pressures (5, 10 and 15 bar) and vapour qualities (0.1, 0.5 and 0.9). These preliminary results suggest that the fundamental derivative remains below unity throughout the two-phase dome under both limiting cases, with the lowest values obtained in close proximity to the critical point and saturated liquid line. The results also suggest that the deviation between the equilibrium and frozen models increases as the reservoir vapour quality increases. Ultimately, this implies that NICFD effects are likely to be present during the a two-phase expansion of MM within novel turbomachinery. However, further research is necessary to investigate these effects under more realistic assumptions that are relevant to turbomachinery applications (*i.e.*, in-homogeneous and non-equilibrium flow).

Keywords: Fundamental derivative · Two-phase flows · Equilibrium flows · Non-equilibrium flows · Organic rankine cycles

1 Introduction

The potential benefit of using two-phase expansion within organic Rankine cycles (ORC) has been widely reported within the literature to enhance performance in waste-heat recovery applications [4]. However, the widespread use of two-phase

M. Pini et al. (Eds.): NICFD 2020, ERCOFTAC Series 28, pp. 3–11, 2021.
https://doi.org/10.1007/978-3-030-69306-0_1

expansion has not yet been realised due to a lack of two-phase expanders. Existing research has focussed on volumetric expanders, such as twin-screw expanders [8], but these expanders are limited in their built-in volume ratio, which limits them to relatively low temperature applications (typically around 100 °C) that correspond to low expansion volume ratios. Comparatively, turbo-expanders are capable of achieving large expansion ratios over a single stage, but are not designed for two-phase expansion. However, it is postulated here that with the use of molecularly complex working fluids, whose saturated-vapour line has a positive gradient when viewed in the temperature-entropy plane, it could be possible to design a single-stage turbine and confine the two-phase expansion region to the stator vane [3]. Preliminary thermodynamic cycle investigations have identified MM as a suitable working fluid.

However, the use of such fluids introduces the possibility of non-ideal compressible-fluid dynamic (NICFD) effects, particularly around the critical point. The topic of two-phase flows of fluids that may exhibit NICFD effects has been studied previously. For example, the work of Thompson [9,10] has explored shocks in retrograde fluids in which there is a phase transition across the shock. More recently, the work Nannan et al. has evaluated the fundamental derivative Γ of two-phase fluids in the near-critical region [6], and attempted to establish the region in which rarefaction shock waves are admissible [7]. Within these studies the two phases are assumed to be in equilibrium, which can be rationalised in close proximity to the critical point where surface tension, and the difference in density of the phases, are reduced.

Whilst it is unlikely that within an ORC system designed for two-phase expansion the turbine inlet conditions will be sufficiently close to the critical point such that non-classical fluid dynamic behaviour, characterised by $\Gamma < 0$, will be present, it is likely that the behaviour of the fluid will be non-ideal, and is likely to be characterised by $0 < \Gamma < 1$. However, to the author's knowledge there have not to date been any significant studies that have attempted to investigate the fundamental derivative within two-phase expansion processes relevant to practical two-phase expansion systems. In such systems, the assumption of both thermodynamic and mechanical equilibrium is likely to be an oversimplification owing to the large change in vapour quality across the expansion process, but the inclusion of in-homogeneity and non-equilibrium effects significantly complicates the analysis. Thus, the aim of this paper is initiate investigations into NICFD effects within the two-phase expansion of MM. This is completed by evaluating the fundamental derivative of a homogeneous two-phase mixture of MM under the two limiting cases of thermal equilibrium and frozen thermal states by first defining expressions for the speed of sound for the two-phase mixture. The two cases of thermal equilibrium and frozen thermal states correspond to two extreme cases where heat transfer between the two phases occurs either instantaneously, or is negligible, and in a realistic application it is rationalised that the speed of sound can be approximated as some combination of the two limiting cases [1].

2 Modelling

For a single-phase fluid, the possibility of NICFD effects can be investigated by evaluating the fundamental derivative, which can be defined as:

$$\Gamma = 1 + \frac{\rho}{a}\left(\frac{\partial a}{\partial \rho}\right)_s, \tag{1}$$

where a is the speed of sound, ρ is the density, and s refers to an isentropic process. For a single-phase fluid, Γ can be computed from the equation of state. However, for a two-phase expansion, the calculation of Γ requires a suitable definition for the speed of sound, which in turn depends on the extent to which the vapour and liquid phases interact.

Under the assumption of a homogeneous two-phase mixture under thermal equilibrium the flow equations reduce to a single mass, energy and momentum equation for the mixture, and it is possible to derive an expression for the speed of sound and thus compute the fundamental derivative. If the analysis is extended to consider the other extreme case in which the two phases remain homogeneous but expand so rapidly that there is no time to exchange heat, and thus the two-phases can be assumed to remain perfectly insulated such that they both undergo independent isentropic expansions, it is also possible to derive an expression for the speed of sound [1]. Within a more realistic two-phase expansion of a homogeneous fluid it is rationalised that the speed of sound is some combination of the two limiting cases. Unfortunately, for the most realistic case in which the flow is neither homogeneous, or under thermal equilibrium, it becomes necessary to solve separate energy and momentum equations for each phase, and the definition for the speed of sound of the two-phase mixture becomes impossible, and the validity of the fundamental derivative questionable.

Within this work, to enable at least a preliminary quantification of the role of NICFD effects within two-phase expansion it is assumed that the two-phase mixture is in a homogeneous state, which implies that the mixture is well mixed, the size of the disperse particles is sufficiently small such that there is no relative motion between the two phases, and thus the speed of sound is not affected by topological features introduced by an in-homogeneous distribution. Under this assumption, the speed of sound can be defined, and Γ evaluated. Whilst the assumption of homogeneity is likely to be an over simplification for a two-phase expansion within a turbomachine where there is likely to be both a large change in pressure and vapour quality, and hence density of each phase, this assumption does permit a preliminary investigation into NICFD effects within two-phase expansions which can be followed up in later studies.

2.1 Homogeneous Equilibrium Modelling

Under the assumption of a homogeneous two-phase mixture, it is assumed that the fluid is well mixed and the size of the disperse particles is sufficiently small such that there is no relative motion between the liquid and vapour phases.

If this is further simplified by neglecting surface tension, then the two-phases can be assumed to be at an equal pressure and travelling at the same velocity (*i.e.*, $p_\mathrm{L} = p_\mathrm{V}$, $u_\mathrm{L} = u_\mathrm{V}$, where p is the pressure and u is the velocity). Similarly, under the assumption of thermal equilibrium the vapour and liquid phases are assumed to transfer heat instantaneously and thus the temperature of both phases is equal to the saturation temperature (*i.e.*, $T_\mathrm{L} = T_\mathrm{V} = T_\mathrm{sat}$, where T is the temperature). Thus, considering the expansion of a homogeneous flow at thermal equilibrium from a defined reservoir condition, defined by the stagnation pressure p_0 and vapour quality X_0, to a specified discharge pressure p, the flow conditions can be calculated, assuming an isentropic expansion:

$$u = \sqrt{2(h_0(p_0, X_0) - h(p, s_0))}\,; \tag{2}$$

$$X = \frac{h(p, s_0) - h_\mathrm{L}}{h_\mathrm{V} - h_\mathrm{L}}\,, \tag{3}$$

where h_0 and s_0 are the enthalpy and entropy at the reservoir conditions, h is the static enthalpy of the two-phase mixture, u is the flow velocity, X is the vapour mass fraction, and 'V' and 'L' refer to the vapour and liquid components respectively.

Under the assumption of homogeneous equilibrium flow, the definition of the fundamental derivative used to investigate the presence of non-ideal compressible-fluid dynamic effects within single-phase fluids (Eq. 1), can, in theory, be extended to two-phase fluids. Thus, the variation in Γ along an isentropic expansion can be determined from the variation in the density and speed of sound, which in turn can be computed based on the known pressure p and mass-fraction X.

The density of the mixture is given as a volume weighted sum of the vapour and liquid components:

$$\rho = \alpha\rho_\mathrm{V} + (1-\alpha)\rho_\mathrm{L}\,, \tag{4}$$

where α is the volume fraction of vapour and is related to the mass fraction as follows:

$$\alpha = \frac{1}{1 + \dfrac{\rho_\mathrm{V}}{\rho_\mathrm{L}}\left(\dfrac{1-X}{X}\right)}\,. \tag{5}$$

Following the analysis described by Brennen [1], the sound of speed for a homogeneous two-phase flow under thermal equilibrium can be found from:

$$\frac{1}{a^2} = \rho\left[\alpha g_\mathrm{V} + (1-\alpha) g_\mathrm{L}\right]\,, \tag{6}$$

where the parameters g_V and g_L are thermodynamic parameters defined by:

$$g_i = \frac{1}{\rho_i}\left(\frac{\partial \rho_i}{\partial p}\right)_e - \frac{1}{h_\mathrm{LV}}\left(\frac{1}{\rho_\mathrm{V}} - \frac{1}{\rho_\mathrm{L}}\right)\left(1 - \rho_i\left(\frac{\partial h_i}{\partial p}\right)_e\right)\,, \tag{7}$$

where the subscript i refers to the phase of interest, h_LV is the latent heat of vapourisation, and the subscript e indicates the phase equilibrium derivative.

Together, Eqs. 6 and 7 are equivalent to the expression reported by Nannan et al. [6], albeit expressed in terms of ρ and h rather than entropy and specific volume. Throughout this work, the multi-parameter Span-Wagner Helmholtz equation of state [2,5] is used to determine all fluid properties. This includes the calculation of metastable properties, although the validity of this approach does require further investigation. The partial derivatives in Eq. 7 are computed using the second-order accurate central difference method with a step size of $\Delta p = 10$ Pa.

Thus, for a given static pressure p and vapour quality X, the saturated liquid and vapour enthalpies and densities can be found, which facilitates the calculation of α, ρ and a through Eqs. 4–7. The derivative $(\partial a/\partial \rho)_s$ is then computed numerically, again using the central difference method with a step size of $\Delta p = 10$ Pa, and the fundamental derivative calculated. To validate the method developed here it has been compared to the results previously obtained for the fundamental derivative of methane [6]. The results from the validation are reported in Fig. 1, although it should be noted that the previous results are obtained using analytical expressions, rather than the numerical approach adopted here.

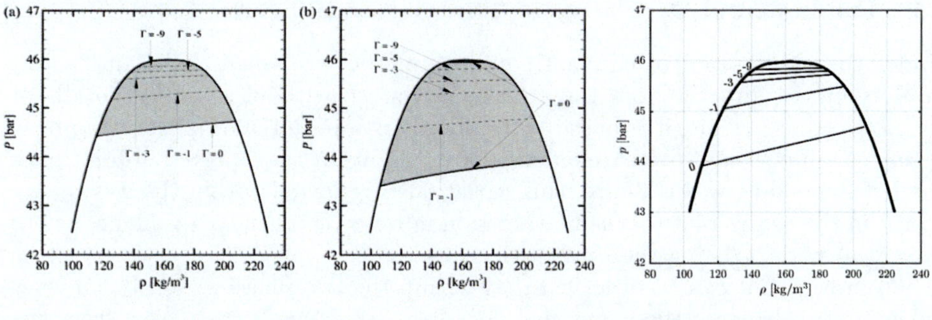

Fig. 1. Calculation of Γ in the critical region of methane. The left and centre plots are taken from Nannan et al. [6] and are computed using both the scaled fundamental equation of state (left) and the Wagner-Setzmann Helmholtz-energy-based equation of state (centre). The right plot is generated using the model developed in the present work.

2.2 Homogeneous Frozen-State Modelling

Following the same rationale as for the previous section, the assumption of a homogeneous fluid leads to $p_L = p_V$ and $u_L = u_V$. However, rather than assume thermal equilibrium between the phases, the frozen-state model assumes that the two-states do not exchange heat or mass and undergo independent isentropic expansions. Thus, based on the defined reservoir conditions (*i.e.*, p_0, X_0), the entropy of the liquid and vapour are obtained from the equation of state, namely $s_L = f(p_0,\ X = 0)$ and $s_V = f(p_0,\ X = 1)$. Then, following an expansion to a defined pressure, p, the density of the liquid and vapour follow from $\rho_L = f(p,\ s_L)$ and $\rho_V = f(p,\ s_V)$. The speed of sound for each phase follow from the thermodynamic definition:

$$a_i = \sqrt{\left(\frac{\partial p}{\partial \rho_i}\right)_s}\,, \tag{8}$$

where again i refers to the phase of interest, s refers to an isentropic process, and the derivative is computed using the central difference method with a step size of 10 Pa. The speed of sound for the two-phase mixture then follows from [1]:

$$\frac{1}{a^2} = \rho\left[\frac{\alpha}{\rho_V a_V^2} + \frac{(1-\alpha)}{\rho_L a_L^2}\right], \tag{9}$$

where ρ and α are obtained using Eqs. 4 and 5 respectively, noting that since there is no mass exchange that the vapour quality remains constant (*i.e.*, $X = X_0$). With the speed of sound and density of the two-phase mixture defined along an isentropic expansion, the derivative $(\partial a/\partial \rho)_s$ is then computed numerically, again using the central difference method with a step size of $\Delta p = 10$ Pa, and the fundamental derivative calculated.

3 Results

3.1 Fundamental Derivative in Thermal Equilibrium

Under the assumption of thermal equilibrium the fundamental derivative can be determined directly from the equation of state without knowing details of the flow process. Thus, it is possible to directly compute Γ within the two-phase region as a function of pressure p and vapour quality X, as reported in Fig. 2. On the left-hand side of this figure, this variation is reported within the T-s plane, whilst in the centre plot this same variation in reported in the p_r-X plane, where $p_r = p/p_c$ with p_c being the fluid critical pressure.

From Fig. 2, it can be observed that within the two-phase region Γ is always below unity, although the value of Γ does increase as one moves away from the critical point, and further away from the saturated liquid region. In a generalised sense, this implies that under the assumption of homogeneous equilibrium a two-phase flow of MM at any combination of pressure and vapour quality may experience NICFD effects following a sudden small change in pressure.

To aid the discussion relating to turbomachinery applications, and specifically two-phase expansions within a stator vane where the expansion is likely to be relatively rapid, three representative expansions from reservoir pressures of 5, 10 and 15 bar, each with reservoir vapour quality of 0.1, are also reported in the right-hand plot of Fig. 2. For these three expansions the pressure ratios corresponding to an isentropic expansion to a completely vapour state are 24.4, 5.5 and 3.0 for $p_0 = 5$, 10 and 15 bar respectively, whilst the corresponding volumetric expansion ratios are 155.5, 29.5 and 12.4. Thus, the large expansion ratios, particularly for the lowest reservoir pressure, certainly warrant some caution on whether the assumption of either homogeneity or thermal equilibrium are valid. However, as the reservoir condition moves closer to the critical point, where the densities of the liquid and vapour phases are more similar, the expansion

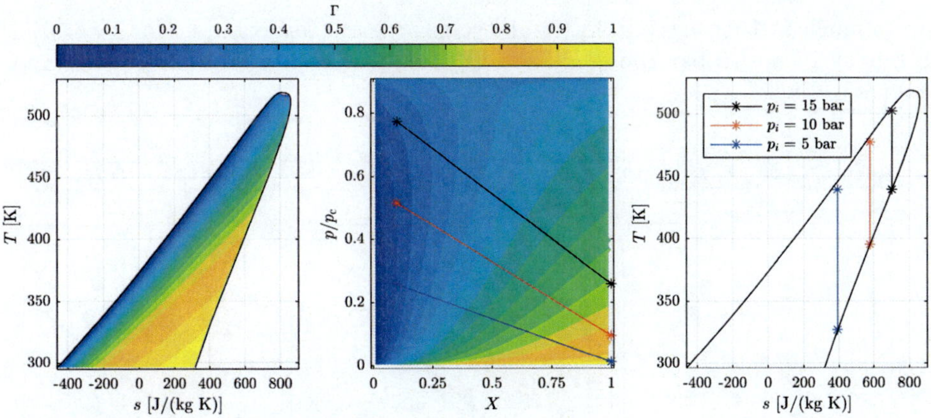

Fig. 2. Variation in Γ within the two-phase region of MM under homogeneous thermal equilibrium reported in the T-s plane (left) and the p-X plane (middle). The right plot reports three isentropic expansions from a reservoir vapour quality of 0.1 and reservoir pressures of 5, 10 and 15 bar with corresponding temperatures of 229.2, 204.1 and 166.4 °C.

ratios are lower and it is easier to rationalise the assumption of homogeneity and thermal equilibrium. Nonetheless, for the three expansions reported, the results indicate that Γ remains below unity for the whole expansion.

3.2 Fundamental Derivative in Frozen and Thermal Equilibrium States

For the three reservoir pressures previously mentioned, a comparison of the variation in Γ during a two-phase expansion of MM from three different reservoir vapour qualities, namely $X_0 = 0.1$, 0.5 and 0.9, under the assumption of a homogeneous mixture in either a frozen or thermal equilibrium state is reported in Fig. 3. Within this figure, the x-axis reports the pressure in non-dimensional form where expansion is considered from the reservoir pressure to a minimum pressure $p_{\min}$ of 1 bar. For the equilibrium cases, the calculation is terminated at the point where the flow becomes saturated vapour.

Considering first the results for the thermal equilibrium cases, it is found that, in line with Fig. 2, Γ increases during the expansion, whilst starting the expansion from a lower reservoir vapour quality leads to lower values for Γ throughout the expansion. The results also reveal that the variation in Γ is more sensitive to the initial reservoir pressure at higher vapour qualities. For example, at $X_0 = 0.9$ the value of Γ in the initial part of the expansion is around 0.5 and 0.7 for reservoir pressures of 15 and 5 bar respectively, whilst for $X_0 = 0.1$ the value of Γ is around 0.25 and 0.2 respectively. This behaviour can be rationalised by considering that for a high pressure of 15 bar and $X_0 = 0.9$, a significant fraction of the flow is high pressure vapour which is in relatively close proximity to the dense-gas region, whilst at 5 bar and $X_0 = 0.9$, the vapour portion of the flow is much further away from the dense-gas region; for example, for saturated vapour at 15 and 5 bar the corresponding compressibility factors are 0.54 and 0.80 respectively.

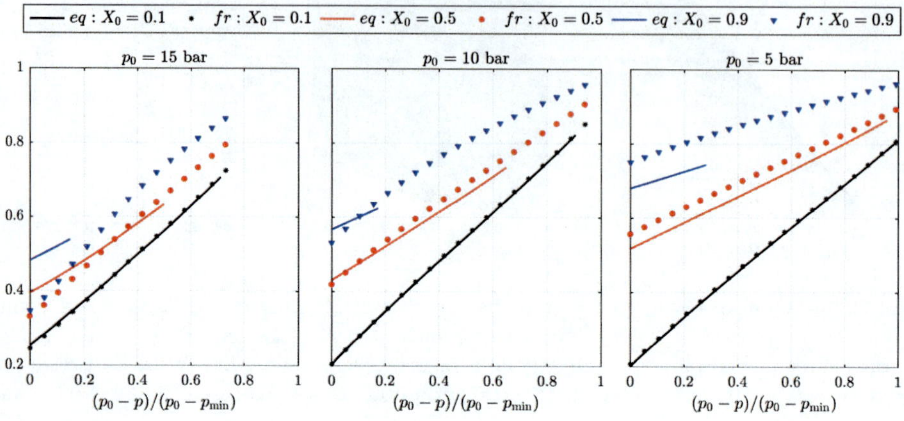

Fig. 3. Variation in Γ for a two-phase expansion of MM from reservoir pressures of 15, 10 and 5 bar and vapour qualities $X_0 = 0.1$, 0.5 and 0.9 under the assumption of frozen (fr) and thermal equilibrium (eq) states.

Comparing now the equilibrium and frozen cases, a number of observations can be made. Firstly, under both assumptions Γ remains below unity throughout the expansion process, indicating that NICFD effects are likely to be present, irrespective of the thermal interaction between the two phases. It can also be rationalised that in a realistic application, where heat transfer neither occurs instantaneously or is negligible, and the speed of sound can be approximated as some combination of the frozen and equilibrium states [1], that NICFD effects will be present. Secondly, it is observed that at low vapour qualities, the deviation between the two extremes appears negligible, but the deviation increases as the initial vapour quality is increased. For high reservoir pressures, the frozen-state assumptions leads to a lower fundamental derivative than the equilibrium assumption during the initial phase of expansion. In comparison, at lower reservoir pressures the frozen-state assumption leads to higher values of the fundamental derivative. The reason for this behaviour requires further investigation.

In summary, the results reported in Fig. 3 suggest that under the assumptions made within this study, and following the definition of the fundamental derivative for single-phase fluids, NICFD effects are likely within the two-phase expansion of MM from a wide range of reservoir pressures and vapour qualities. This finding holds under both assumptions for the thermal exchange between the two phases. Moreover, the deviation between the frozen-state and thermal equilibrium assumptions increases at higher vapour qualities and appears most significant at high reservoir pressures. In the context of a two-phase ORC turbine, where an optimal expansion would likely begin from a relatively high reservoir pressure and low vapour quality, these preliminary results seem to suggest that the degree to which NICFD effects are present is not affected by the amount of heat transfer between the two-phases. However, this finding requires further interrogation.

4 Conclusions and Future Works

This paper has detailed preliminary investigations into expanding the concept of the fundamental derivative to evaluating NICFD effects within the two-phase expansion of MM under operating conditions relevant to an ORC power system operating with two-phase expansion. For the assumption of a homogeneous two-phase mixture under the two limiting cases of thermal equilibrium and frozen thermal states, expressions for the speed of sound are defined and these are used to evaluate the fundamental derivative. The results have revealed that under both cases the expansion of MM from reservoir pressures of 5, 10 and 15 bar with reservoir vapour qualities of 0.1, 0.5 and 0.9 remains within a region where $0 < \Gamma < 1$ indicating the presence of NICFD effects. Moreover, it is found that the deviation in the fundamental derivative predicted under the frozen-state and thermal equilibrium assumptions is increased at higher vapour qualities and appears most significant at high reservoir pressures. However, it should be noted that the assumption of a homogeneous two-phase mixture is likely to be an oversimplification, and further investigations are required to fully understand the role of the fundamental derivative within two-phase expansion.

References

1. Brennen, C.: Fundamentals of Multiphase Flows. Cambridge University Press, Cambridge (2005)
2. Colonna, P., Nannan, N., Guardone, A., Lemmon, E.: Multiparameter equations of state for selected siloxanes. Fluid Phase Equilib. **244**, 193–211 (2006)
3. Elliott, D.: Theory and tests of two-phase turbines. Technical report, U.S. DoE Technical report: 10614-1, California, U.S. (1982)
4. Fischer, J.: Comparison of trilateral cycles and organic Rankine cycles. Energy **36**, 6208–6219 (2011)
5. Lemmon, E., Huber, M., McLinden, M.: NIST standard reference database 23: Reference fluid thermodynamic and transport properties-REFPROP (2013)
6. Nannan, N.R., Guardone, A., Colonna, P.: On the fundamental derivative of gas dynamics in the vapor-liquid critical region of single-component typical fluids. Fluid Phase Equilib. **337**, 259–273 (2013)
7. Nannan, N.R., Sirianni, C., Mathijssen, T., Guardone, A., Colonna, P.: The admissibility domain of rarefaction shock waves in the near-critical vapour-liquid equilibrium region of pure typical fluids. J. Fluid Mech. **795**, 241–261 (2016)
8. Read, M.G., Smith, I.K., Stosic, N.: Optimisation of power generation cycles using saturated liquid expansion to maximise heat recovery. Proc. Inst. Mech. Eng. E-J. Process **23**, 57–69 (2017)
9. Thompson, P.A., Carofano, G., Kim, Y.: Shock waves and phase changes in a large-heat-capacity fluid emerging from a tube. J. Fluid Mech. **166**, 57–92 (1986)
10. Thompson, P.A., Sullivan, D.A.: On the possibility of complete condensation shock waves in retrograde fluids. J. Fluid Mech. **70**, 639–649 (1975)

Non-ideal Gas Effects on Supersonic-Nozzle Transfer Functions

Stephen D. Winn[1,2(✉)] and Emile Touber[1,2]

[1] Department of Mechanical Engineering, Imperial College London, London, UK
stephen.winn16@imperial.ac.uk
[2] Shocks, Solitons and Turbulence Unit, Okinawa Institute of Science and Technology, Okinawa, Japan
emile.touber@oist.jp

Abstract. Flow disturbances traveling in a nozzle can be amplified or attenuated and generate excessive acoustic noise (e.g. jet engine exhaust) or interact with shocks to cause excessive loading on components (e.g. between turbine blades). Non-ideal gas-dynamic effects are investigated within the framework of linearised inviscid quasi-one dimensional nozzle flow. The transfer function of choked supersonic divergents is investigated when prescribing an inlet entropy perturbation. Initial results using a van der Waals gas highlight the contrast with ideal gas behaviour with and without a shock in the divergent. For the chosen conditions, in the shock-free configuration, a five-fold increase in amplification of pressure perturbations at higher wavelengths (relative to nozzle length) and stronger attenuation (over one order of magnitude lower) at lower wavelengths is observed when compared to an ideal gas. In the shocked configuration, greater amplification is again observed in the van der Waals case owing to the selectivity of the shock in amplifying the incoming density perturbations. Furthermore, up to an order of magnitude greater shock displacement is observed over the range of perturbation wavelengths in the van der Waals case.

Keywords: Transfer function · Quasi-one dimensional nozzle flow · Non-ideal gas effects · Thermodynamic critical point · Dense gas

1 Linearised Quasi-One Dimensional Nozzle Flow

The equations governing the flow herein are the inviscid, quasi-one dimensional Euler equations:

$$\frac{\partial}{\partial t}\begin{bmatrix}\rho\\ \rho u\\ \rho e_t\end{bmatrix} + \frac{\partial}{\partial x}\begin{bmatrix}\rho u\\ \rho u^2+p\\ u(\rho e_t+p)\end{bmatrix} + \frac{1}{A}\frac{\mathrm{d}A}{\mathrm{d}x}\begin{bmatrix}\rho u\\ \rho u^2\\ u(\rho e_t+p)\end{bmatrix} = \begin{bmatrix}0\\0\\0\end{bmatrix}, \tag{1}$$

where ρ, u, p, e_t are, respectively, the fluid density, velocity, pressure, total energy (where $e_t = e + u^2/2$) and A is the nozzle cross-section area. These are supplemented with the equation of state for internal energy, e, for an arbitrary gas.

M. Pini et al. (Eds.): NICFD 2020, ERCOFTAC Series 28, pp. 12–19, 2021.
https://doi.org/10.1007/978-3-030-69306-0_2

An infinitesimal harmonic (in time) perturbation of the base-flow is considered for any of the primitive variables $q \in \{\rho, u, p\}$ such that:

$$q(x,t) = \bar{q}(x) + \epsilon q(x)' e^{i\omega t}, \tag{2}$$

where ω is the frequency of the perturbation ($\omega \in \mathbb{R}_{>0}$), ϵ is an arbitrarily-small non-dimensional parameter ($\epsilon \ll 1$) and $i^2 = -1$. Injecting this form in the governing equations and retaining only first order terms yields the linearised equations without any assumption on the gas model of the form:

$$\begin{bmatrix} \bar{u} & \bar{\rho} & 0 \\ 0 & \bar{u} & \frac{1}{\bar{\rho}} \\ 0 & \bar{\rho} c^2 & \bar{u} \end{bmatrix} \frac{\mathrm{d}\mathbf{q}'}{\mathrm{d}x} = \left(\frac{1}{(\bar{M}^2 - 1)A} \frac{\mathrm{d}A}{\mathrm{d}x} \begin{bmatrix} -\tilde{c}\bar{M}^3 & \bar{\rho} & 0 \\ \frac{\bar{M}^2 \tilde{c}^2}{\bar{\rho}} & -\bar{M}\tilde{c} & 0 \\ -\bar{M}^3 \tilde{c}^2 \left(\tilde{c} + 2\bar{\rho}\tilde{c}_\rho\right) & \bar{\rho}\tilde{c}^2 & -2\bar{M}^3 \tilde{c}^2 \tilde{c}_p \bar{\rho} \end{bmatrix} - i\omega \mathbf{I} \right) \mathbf{q}', \tag{3}$$

where $\mathbf{q}' = [\rho', u', p']^T$ is the vector of complex perturbation values, $\tilde{c} = c(\bar{\rho}, \bar{p})$ is the isentropic speed of sound, c, evaluated on the base-flow thermodynamic properties and $\tilde{c}_\rho = (\partial c/\partial \rho)_p (\bar{\rho}, \bar{p})$, $\tilde{c}_p = (\partial c/\partial p)_\rho (\bar{\rho}, \bar{p})$. These are supplemented with the Rankine–Hugoniot jump conditions and their linearised counter-part to relate fluctuations either side of a shock to the base-flow values and their gradients:

$$\begin{bmatrix} \bar{u}_b & \bar{\rho}_b & 0 \\ \bar{u}_b^2 & 2\bar{\rho}_b \bar{u}_b & 1 \\ -\left.\frac{\partial \wp}{\partial \rho}\right|_b & 0 & 1 \end{bmatrix} \begin{bmatrix} \rho_b' \\ u_b' \\ p_b' \end{bmatrix} + \begin{bmatrix} -i\omega [\![\bar{\rho}]\!] \\ \frac{\bar{\rho}_a^2 \bar{u}_a^2}{A} \frac{\mathrm{d}A}{\mathrm{d}x} \left[\!\left[\frac{1}{\bar{\rho}}\right]\!\right] \\ \left.\left(\frac{\partial \wp}{\partial p} \frac{\mathrm{d}\bar{p}}{\mathrm{d}x} + \frac{\partial \wp}{\partial \rho} \frac{\mathrm{d}\bar{\rho}}{\mathrm{d}x} \right)\right|_a + \left.\left(\frac{\partial \wp}{\partial \rho} \frac{\mathrm{d}\bar{\rho}}{\mathrm{d}x} - \frac{\mathrm{d}\bar{p}}{\mathrm{d}x} \right)\right|_b \end{bmatrix} x_s' = \begin{bmatrix} \bar{u}_a & \bar{\rho}_a & 0 \\ \bar{u}_a^2 & 2\bar{\rho}_a \bar{u}_a & 1 \\ \left.\frac{\partial \wp}{\partial \rho}\right|_a & 0 & \left.\frac{\partial \wp}{\partial p}\right|_a \end{bmatrix} \begin{bmatrix} \rho_a' \\ u_a' \\ p_a' \end{bmatrix}, \tag{4}$$

where x_s' is the shock displacement and $\wp = \wp(\rho, p, M)$ is the shock-adiabat pressure (such that $p_b = \wp(\rho_a, p_a, M_a)$), the derivatives of which are evaluated using the base-flow values at the locations shown in Fig. 2. Integrating Eq. 3 yields the spatial distribution of $\mathbf{q}'$ in a nozzle and thus the transfer function (modulus and phase at the outlet) resulting from any combinations of physical boundary conditions. A divergent with a linear increase in cross-section area (henceforth referred to as a linear nozzle) is employed to expand a gas from an initial strictly supersonic state (see schematic Fig. 3, left). Two cases are investigated: a shock free nozzle and a divergent containing a compression shock. It is assumed that the cross-section area is constant upstream and downstream of the linear nozzle. An inlet perturbation in the form of entropy and acoustic waves is prescribed. Such an analysis has been used for an ideal gas to provide useful insight into the transfer function of choked nozzles e.g. [2].

2 Non-ideal Gas Model

Non-ideal gas properties are explored herein using the van der Waals (vdW) gas equation. In the vicinity of the thermodynamic critical point (TCP), dense gases (i.e. gases that are characterised by a large ratio of specific heat, c_v to gas constant, R) present a region where the fundamental derivative of gas dynamics,

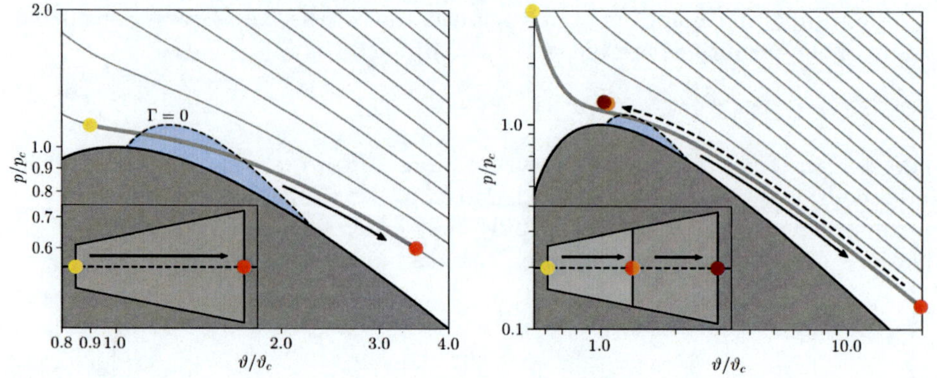

Fig. 1. (p-ϑ) diagrams for a van der Waals gas (with $R/c_v = 1.3 \times 10^{-2}$). Isentropes are shown in thin grey lines, with the region where they become concave ($\Gamma < 0$) shaded in light blue. The two-phased region is shown in dark-grey. Left: isentrope chosen for the supersonic expansion of Sect. 3.1. Right: supersonic expansion isentrope (yellow → red) and subsonic compression (orange → maroon) chosen for Sect. 3.2. A schematic of the corresponding locations in the nozzle is given in the bottom left.

$\Gamma \equiv 1 + (\rho/c)\left(\partial c/\partial \rho\right)_s$, can become negative (see Fig. 1) denoting locally concave isentropes. Admissible steady-state flows in this region have been explored and categorised (see e.g. [4]). For supersonic expansions close to the TCP, certain stagnation conditions lead to the Mach number having a non-monotonous behaviour as shown in Fig. 3 (left) which is impossible in ideal gases (we restrict ourselves to expansions that contain only one sonic point). Furthermore, the energy transfer properties of shocks in this region present an acute sensitivity to the upstream state and Mach number, explored in [1], and allow for a degree of control over the energy redistribution downstream of the shock. In Sects. 3 and 4, the computations are carried out using the van der Waals gas model and the ideal gas model for comparison. In both cases, the ratio $R/c_v = 0.013$ is used which is representative of PP10 gas [1].

3 Transfer Function of a Choked Supersonic Nozzle

The results for the transfer function of the supersonic nozzle is divided into two steps as illustrated in Fig. 2. First, results for the transfer function of a supersonic expansion alone are presented followed by the results with the addition of a shock in the divergent section.

3.1 Transfer Function of a Shock-Free Nozzle

For given thermodynamic inlet (subscript 1) and outlet states (subscript 2), the transfer function of a linear nozzle for a given inlet perturbation depends on two parameters: the mass flow through the nozzle, $\dot{m}$, and the ratio of the perturbation wavelength, λ, and the nozzle length, L. Figure 3 presents the transfer

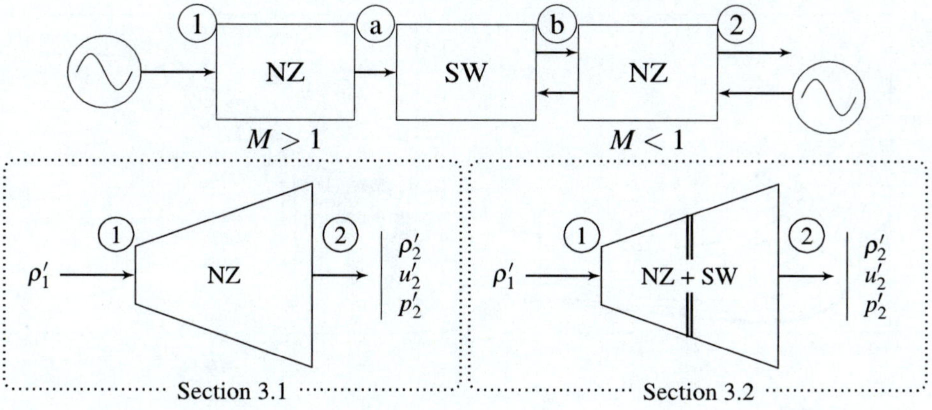

Fig. 2. Top: Schematic of the 'blocks' in the transfer function of a choked supersonic nozzle (NZ) containing a shock (SW). Bottom: shock-free and shocked nozzle 'blocks' with notations for the two configurations considered.

functions (in terms of modulus) for an inlet entropy perturbation over a range of λ/L values. The inlet perturbation generates greater ($\times 5$) pressure perturbation in the vdW case for larger wavelengths (relative to nozzle length). This is of interest, for example, when considering 'indirect noise' generation in turbines (i.e. acoustic noise generated from entropy perturbations [3]). Furthermore, at higher wavelength, the density perturbations undergo greater amplification in the vdW case than in the ideal gas case which could exacerbate the shock transfer effects discussed in following section (and [1]). However, for $\lambda \leq L$, the density and pressure perturbations are, over most of the range, attenuated in the vdW case compared to the ideal gas (e.g. at $\lambda/L = 0.1$, $|p'_{2,vdw}|/|p'_{2,ig}| \approx 0.2$ and $|\rho'_{2,vdw}|/|\rho'_{2,ig}| \approx 9 \times 10^{-2}$).

3.2 Transfer Function of a Shocked Nozzle

The addition of a shock in the nozzle requires specifying two additional boundary conditions: one is the exit thermodynamic state (state 2 in Fig. 2, bottom right) to determine where the post-shock subsonic compression ends. The other is the perturbation boundary conditions i.e. the magnitude and phase of upstream propagating acoustic wave at the exit of the nozzle. In the case presented here, there is no back-propagating forcing imposed at the exit of the nozzle. Furthermore, for the following results, the shock is placed at $\bar{x}_s = 2L/3$ thus determining the extent of the subsonic compression and the exit conditions. The shock configuration presented in this section was chosen based on the work carried out by [1]. In the isolated shock case, Alferez & Touber identified shock transmission properties allowing for a strong amplification of the entropy mode (up to two orders of magnitude greater than the ideal gas case). In the current configuration, presented in Fig. 4 (left), this overall amplification difference through the

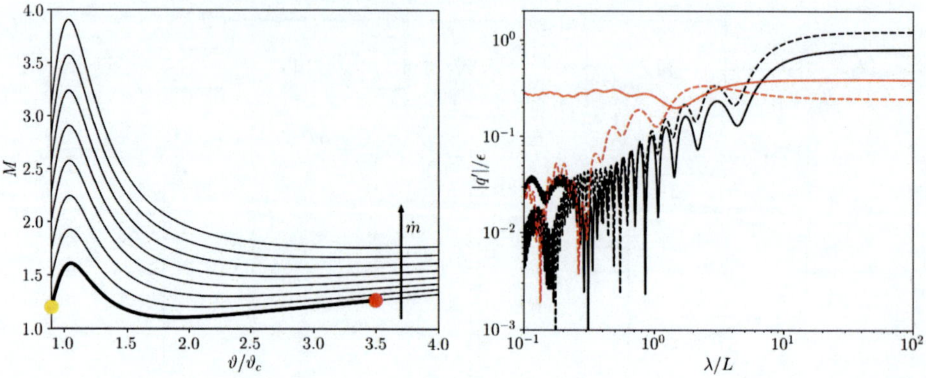

Fig. 3. Left: evolution of the Mach number along the thicker isentrope of Fig. 1 (left) ($s_0 = s(\bar{T}_0, \bar{p}_0)$, where $\bar{T}_0 = T_c$ and $\bar{p}_0 = 0.6p_c$) for various mass flows. The thicker line denotes $\dot{m}$ used in the shock-free case. Subscript c denotes critical point values. Right: Transfer function (modulus only) of the supersonic expansion along the thick isentrope given in Fig. 1 (left) with $M_1 = 1.2$ for an inlet entropy perturbation for various wavelengths. Solid and dashed lines are for $|\rho'_2/\epsilon|$ and $|p'_2/\epsilon|$ respectively, with red for ideal gas, black for vdW gas.

nozzle is comparable (the maximum ratio $|\rho'_{2,vdw}|/|\rho'_{2,ig}| \approx 149$ for $\lambda/L \approx 0.14$). For both ideal gas and vdW gases, the transfer function for the density perturbation exhibits relatively little sensitivity to the perturbation wavelength for higher wavelength and becomes more sensitive when $\lambda/L < 1$. For the pressure perturbation however, there exists a wavelength which allows for a local minimum in the perturbation value for the vdW gas. The shock transfer properties are not only affected by the non-ideal effects linked to the proximity to the thermodynamic critical point but also by the non-uniformity of the base-flow either side of the shock. The shock displacement, shown in Fig. 4 (right), varies significantly between ideal and vdW gases. The shock displacement in the vdW case is greater than that in the ideal gas case over most of the range of wavelengths presented with $|x'_{s,vdw}|/|x'_{s,ig}| \approx 10$ at $\lambda/L = 100$. The vdW shock displacement features a strong minimum around $\lambda/L \approx 0.15$ causing the magnitude to drop below that of the ideal gas. Transfer functions with slightly different mass flows through the nozzle (maximum $\Delta\dot{m}/\dot{m}_{M_1=1.1} \approx 9\%$ for $M_1 = 1.2$) are included in Fig. 4 to demonstrate the sensitivity of the results to this parameter. The pressure perturbation transfer function is the quantity that presents the strongest sensitivity to mass flux around the point where $|p'_2|$ undergoes a local minimum.

4 Comparison to DNS

In this section, we compare the results obtained via linear analysis to those obtained by solving the non-linear equations when the perturbation amplitude is small (referred to as DNS results). An example case for the vdW gas is chosen

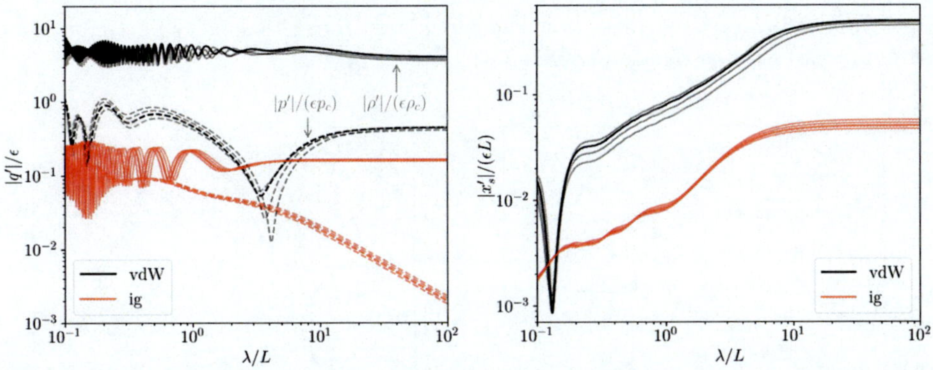

Fig. 4. (Left) modulus of the transfer function where solid and dashed lines are for $|\rho_2'/\epsilon|$ and $|p_2'/\epsilon|$ respectively and (Right) shock displacement magnitude for various length-scale ratios (non-reflective outlet and no outlet forcing) in red for ideal gas, black for vdW gas. The pre-shock conditions are: $p_a/p_c = 0.13$, $T_a/T_c = 1$ and $M_a = 3.1$ (obtained for $M_1 = 1.1$). The exit conditions are $p_2/p_c \approx 1.30$ for the ideal gas case and $p_2/p_c \approx 1.29$ for the vdW gas case. Lightly transparent curves are for slightly different mass flows ($M_1 \in \{1.05, 1.15, 1.2\}$) to illustrate sensitivity of the results to this parameter.

for comparison: for $\gamma = 1.013$, the upstream shock values are set to $M_a = 3.1$, $p_a/p_c = 0.13$ and $T_a/T_c = 1$ for an inlet Mach number $M_1 = 1.1$ i.e. the case of Fig. 4. The non-linear solver integrates Eq. 1 in time. First, a base-flow $\mathbf{q}_b$ is converged in time to a chosen precision. Then, an inlet density perturbation is prescribed at a given frequency ω (to match the desired λ/L value) and a sufficiently small amplitude ϵ (taken to be $\epsilon = 4 \times 10^{-3}$). The solution is advanced in time until a harmonic state is achieved giving the instantaneous perturbed flow denoted by $\mathbf{q}_n(t)$. For the following results, the comparison is carried out from $t = 10\tau$ (where τ is the period of the inlet perturbation) when the behaviour is expected to be harmonic. The numerical results plotted in Fig. 5 are thus $\mathbf{q}_n(t) - \mathbf{q}_b$ for $t \in [10\tau, 11\tau]$. The numerical integration is carried out using a 3rd order Runge–Kutta scheme for time integration and 4th order, 5 point-stencil centered finite difference scheme to calculate spatial derivatives. The computational space is discretised using $n_x = 5.6 \times 10^4$ points in order to resolve the shock displacement (the required number of points being estimated using the linear results and verified *a posteriori*). A constant and uniform artificial bulk viscosity is used to capture the discontinuity and tuned to ensure a scale separation between shock-thickness and perturbation wavelength while simultaneously damping the spurious oscillations at the shock (an approach used and validated in e.g. [1,7]). The inlet and outlet boundary conditions are enforced using a characteristic formulation for an arbitrary gas (see e.g. [5]). The results of Fig. 5 demonstrate the validity of the analysis in the linear regime – here for $\epsilon \sim 10^{-3}$ – as both the magnitude and phase of the shock displacement as well as the spatial distribution of ρ' are accurately recovered. The spatial distribution of the density perturbation illustrates how the shock amplifies the density perturbation and compresses its associated wavelength.

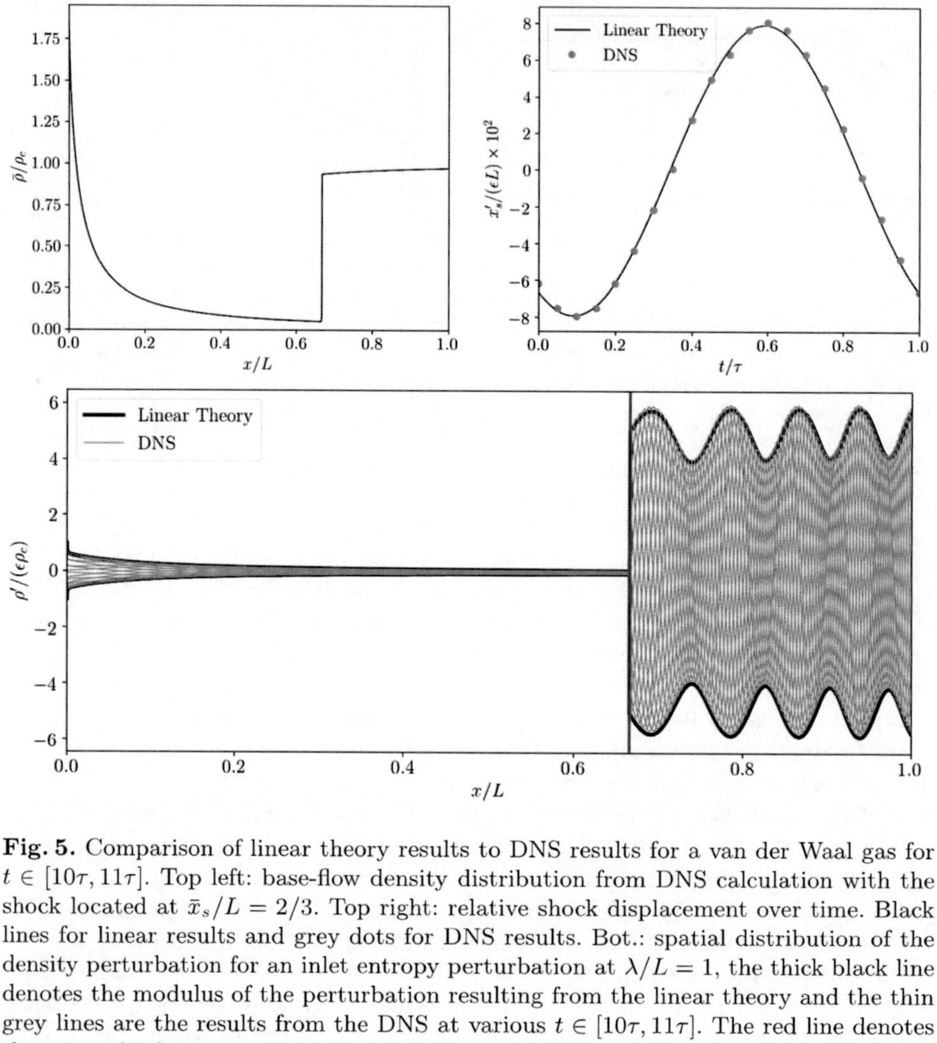

Fig. 5. Comparison of linear theory results to DNS results for a van der Waal gas for $t \in [10\tau, 11\tau]$. Top left: base-flow density distribution from DNS calculation with the shock located at $\bar{x}_s/L = 2/3$. Top right: relative shock displacement over time. Black lines for linear results and grey dots for DNS results. Bot.: spatial distribution of the density perturbation for an inlet entropy perturbation at $\lambda/L = 1$, the thick black line denotes the modulus of the perturbation resulting from the linear theory and the thin grey lines are the results from the DNS at various $t \in [10\tau, 11\tau]$. The red line denotes the mean shock position.

5 Conclusion

Initial results concerning non-ideal effects in a both shock-free and shocked supersonic diffuser have been presented here. In the shock-free case, integration of the linearised equations revealed a different redistribution of the inlet density perturbation across the primitive variables in the vdW case compared to the ideal gas case allowing for a stronger pressure perturbation at the outlet of the nozzle at large wavelengths (compared to the nozzle length). This illustrates that the expansion through the nozzle itself, for a change of equation of state, is a first means of redistribution of the initial inlet perturbation across the primitive variables when compared to an ideal gas. For the case of shocked flow, the shock chosen in Sect. 3.2 further illustrates the importance of the equation of state in the redistribution of incoming primitive variable perturbations as the density perturbation is selectively amplified through this shock. Furthermore, for the conditions chosen in Sect. 3.2, the shock displacement is, over most of the wavelength range, an order of magnitude greater in the vdW case than in the ideal gas case. Such an increase in the displacement could lead to greater aerothermal loads on the divergent (e.g. in the case of shocks between turbine blades operating in a comparable region of (p-ϑ)-space such as those of [6]). The impact of the outlet condition and possible upstream propagating acoustic waves have yet to be explored. For the latter, a possible resonance (or anti-resonance) of the shock displacement could be achieved by tuning the phase of an imposed upstream propagating acoustic forcing at the outlet.

References

1. Alferez, N., Touber, E.: One-dimensional refraction properties of compression shocks in non-ideal gases. J. Fluid Mech. **814**, 185–221 (2017)
2. Duran, I., Moreau, S.: Solution of the quasi-one-dimensional linearized Euler equations using flow invariants and the Magnus expansion. J. Fluid Mech. **723**, 190–231 (2013)
3. Goh, C.S., Morgans, A.S.: Phase prediction of the response of choked nozzles to entropy and acoustic disturbances. J. Sound Vibr. **330**(21), 5184–5198 (2011)
4. Guardone, A., Vimercati, D.: Exact solutions to non-classical steady nozzle flows of Bethe-Zel'dovich-Thompson fluids. J. Fluid Mech. **800**, 278–306 (2016)
5. Okong'o, N., Bellan, J.: Consistent boundary conditions for multicomponent real gas mixtures based on characteristic waves. J. Comput. Phys. **176**(2), 330–344 (2002)
6. Romei, A., Vimercati, D., Persico, G., Guardone, A.: Non-ideal compressible flows in supersonic turbine cascades. J. Fluid Mech. **882** (2020)
7. Touber, E., Alferez, N.: Shock-induced energy conversion of entropy in non-ideal fluids. J. Fluid Mech. **864**, 807–847 (2019)

Check for updates

Formation of Rarefaction Shockwaves in Non-ideal Gases with Temperature Gradients

Nitish B. Chandrasekaran(✉), Bertrand Mercier, and Piero Colonna

Propulsion & Power Group, Aerospace Engineering, TU Delft, Delft, The Netherlands
{n.b.chandrasekaran,b.h.mercier,p.colonna}@tudelft.nl

Abstract. The nonlinear propagation of finite amplitude waves in the non-ideal compressible fluid dynamic (NICFD) region of high-molecular weight fluids, and in particular in cases where non-classical gas dynamic phenomena like the formation of rarefaction shock waves may be expected, is of great scientific interest. Almost all the theoretical developments so far are based on the assumption that the waves propagate in a fluid kept in homogeneous conditions. Experimental activities performed with the Flexible Asymmetric Shock Tube (FAST) operated in the laboratories of the Propulsion and Power group at the Delft University of Technology have shown that obtaining such conditions is particularly challenging, especially keeping the tube at constant temperature. Stemming from this observation, this study is a part of a theoretical and numerical investigations aimed at assessing the influence of temperature gradients at the boundary on wave propagation in so-called Bethe-Zel'dovich Thompson (BZT) fluids. The full-wave Westervelt Equation is solved numerically using the Finite Difference Time Domain (FDTD) method. The steepening of the wave front is used as an indicator of shock formation. The effect of varying temperature, and the corresponding variation of the fundamental derivative of gas dynamics on the distortion of rarefaction waves is analyzed by observing the simulated behaviour of the wave as it propagates from a region of negative values of the fundamental derivative of gas dynamics to a region of positive values.

Keywords: NICFD · BZT · Γ · Non classical gas dynamics · Rarefaction shockwaves · Westervelt equation · Nonlinear distortion

1 Introduction

Non-ideal compressible fluid dynamics (NICFD), and non-classical fluid dynamics within it, have been rather extensively studied theoretically, numerically and experimentally. Most of the studies on non-linear wave propagation in the NICFD regime are based on the assumption that the medium in which the waves travel is in homogeneous conditions. All the so-far unsuccessful experimental efforts aimed at generating and measuring the formation and propagation of rarefaction shock waves in the dense vapor of BZT fluids [1,4,9] also relied and rely on trying to obtain such conditions in order to facilitate the interpretation of experimental results. However, obtaining a sufficiently constant temperature at the surfaces of the volume containing the fluid is challenging. For this reason, but also and more in general, to expand the theory of non-linear

M. Pini et al. (Eds.): NICFD 2020, ERCOFTAC Series 28, pp. 20–25, 2021.
https://doi.org/10.1007/978-3-030-69306-0_3

wave propagation in the NICFD regime, it is of interest to extend the modeling of wave propagation in non-homogeneous fluids to the case of dense vapors of fluids exhibiting negative values of the fundamental derivative of gas dynamics Γ for the thermodynamic states of the fluid affected by the process.

Γ is very sensitive to variations in temperature, therefore a small change in the temperature of the medium can give rise to a completely different nonlinear behaviour of the propagation of the waves. Such propagation is affected by strong effects on the steepening of a nonlinear distortion as the wave progresses through regions of opposite nonlinearities.

An important parameter for the experimental study of nonlinear wave propagation is the shock formation distance. For a uniform medium, this can be computed easily using the method of characteristics (MoC) [2,9]. In this case, the characteristic lines are straight with constant slopes and their intersection indicates shock formation. The MoC is both more complex to solve and exhibits some limitations when the temperature of the fluid varies, in which case the characteristic lines are curved due to the changing value of the local sound speed. In such situations, a more general approach to describe the nonlinear evolution of a disturbance needs to be employed.

The nonlinear steepening of waves in a medium with an axial variation in temperature, and thus a variation in Γ, is investigated using the full-wave Westervelt Equation. The Westervelt equation is a well-known mathematical model in nonlinear acoustics, widely used in several industrial and medical applications including diagnostic ultrasound, sonochemistry, etc. [11]. This equation is solved numerically using the Finite Difference in Time Domain (FDTD) method for the case of a non-isothermal attenuating medium. Various forms of initial disturbances are considered and the steepening of the wave front and the corresponding shock formation distances and times are computed.

2 Westervelt Equation

The Westervelt equation describes the propagation of an acoustic wave in a quiescent medium in the presence of viscous dissipation and nonlinearity effects. In the absence of temperature gradients, simpler equations such as the inviscid Burgers' equation $u_t + cu_x = 0$, where c is the nonlinear sound propagation velocity, can also be employed [7]. However, this equation allows for wave propagation in only one direction. Hence, it is not suitable for studying wave propagations in temperature inhomogeneous mediums wherein the change in the acoustic impedance arising from the variation in temperature reflects a part of the incident wave. The Westervelt equation [6] is written as:

$$\rho_0 \nabla \cdot (\rho_0 \nabla p) + \frac{1}{c_0^2}\frac{\partial^2 p}{\partial t^2} + \frac{\delta}{c_0^4}\frac{\partial^3 p}{\partial t^3} + \frac{\Gamma}{\rho_0 c_0^4}\frac{\partial^2 p^2}{\partial t^2} = 0, \tag{1}$$

where c_0 and ρ_0 are the local sound speed and density of the medium at rest, δ is the dissipation coefficient and Γ, also referred to as the nonlinearity coefficient β in acoustics, is the fundamental derivative of gasdynamics [9]:

$$\Gamma = 1 + \frac{\rho}{c}\left(\frac{\partial c}{\partial \rho}\right)_s \tag{2}$$

The diffusivity coefficient is given by

$$\delta = \frac{1}{\rho_0}\Big(\frac{4}{3}\mu + \mu_B\Big) + \frac{\kappa}{\rho_0}\Big(\frac{1}{c_v} - \frac{1}{c_p}\Big), \tag{3}$$

where μ is the shear viscosity, μ_B is the bulk viscosity, κ is the thermal conductivity and c_p and c_v are the specific heats at constant pressure and volume, respectively [12].

2.1 Numerical Method

Spatial and temporal derivatives in Eq. (1) are approximated with discrete differences using the Finite Difference Time Domain (FDTD) method [5]. The spatial dimension is divided into N_x elements with equal spacing of Δx and is indexed with i. Absorbing boundary conditions are implemented at the domain boundaries to prevent numerical reflections from affecting the results. The spatial derivative is computed using second-order accurate, central differences as

$$\frac{\partial^2 p}{\partial x^2} \approx \frac{1}{(\Delta x)^2}(p_{i+1}^n - 2p_i^n + p_{i-1}^n). \tag{4}$$

The temporal dimension is discretized into N_t timesteps with a uniform spacing of Δt and index n. The time derivatives are computed using second-order accurate, central differences with the nonlinear and absorption terms expanded using backward differencing, resulting in

$$\frac{\partial^2 p}{\partial t^2} \approx \frac{1}{(\Delta t)^2}(p_i^{n+1} - 2p_i^n + p_i^{n-1}), \tag{5a}$$

$$\frac{\partial^3 p}{\partial t^3} \approx \frac{1}{2\Delta t^3}(5p_i^n - 18p_i^{n-1} + 24p_i^{n-2} - 14p_i^{n-3} + 3p_i^{n-4}), \tag{5b}$$

$$\frac{\partial^2 p^2}{\partial t^2} = 2\Big[\Big(\frac{\partial p}{\partial t}\Big)^2 + p\frac{\partial^2 p}{\partial t^2}\Big] \approx 2\Big[\frac{3p_i^n - 4p_i^{n-1} + p_i^{n-2}}{2\Delta t} + \frac{2p_i^n - 5p_i^{n-1} + 4p_i^{n-2} - p_i^{n-3}}{\Delta t^2}\Big]. \tag{5c}$$

Equations (5) are substituted into Eq. (1) and solved for p_i^{n+1}. Since Eq. (5b) requires five time steps for initialization, the starting waveform is advected using the second-order approximation for nonlinear propagation velocity for $n = 1 \dots 5$.

Computations are performed in a domain of length 9 m with a uniform grid of 1 mm spatial discretization. The time step of the simulation is chosen to be 1 μs to ensure numerical stability. The source is simulated to be siloxane D_6, a BZT fluid, whose medium temperature and pressure are set at 369 °C and 9 bar respectively. The value of Γ estimated with technical equations of state based on Helmholtz energy [3,8] at this state is –0.15, therefore the flow is theoretically affected by nonclassical gas dynamic conditions. The dimensions of the numerical domain and the initial conditions of the simulation are chosen to match the Flexible Asymmetric Shock Tube (FAST) facility, a shock-tube test setup designed at the Delft University of Technology with the aim of studying the propagation of expansion waves in the NICFD regime [9].

The location at which a shock wave forms is identified by plotting the inverse of the slope of the wave front along the x-direction. Since the shock wave is a discontinuity with the slope of the wave front being infinite, the point at which the inverse of the slope becomes zero provides the shock formation distance [10]. The slope here is expressed as the gradient in the local acoustic velocity $(\partial u/\partial x)$ along the wavefront. To prevent numerical instabilities, the domain is defined to have $\Gamma = 0$ till $0.2 \cdot L_0$ after which Γ is calculated using the thermodynamic model.

3 Results

Figure 1 illustrates a result of simulating the propagation of a right-travelling triangular rarefaction wave of amplitude 50 kPa and wavelength 0.72 m in D_6 with uniform temperature. The nonlinear steepening is shown in Fig. 1 (a) by comparing the waveform of the same initial disturbance which travels in a hypothetical linear medium, and in a non-linear medium. Since the fluid temperature is constant, the value of $(\partial u/\partial x)^{-1}$ is expected to decrease linearly down to zero. Figure 1 (b) illustrates this trend, though calculated values are also affected by a spurious oscillation which is expected to occur when estimating the slope.

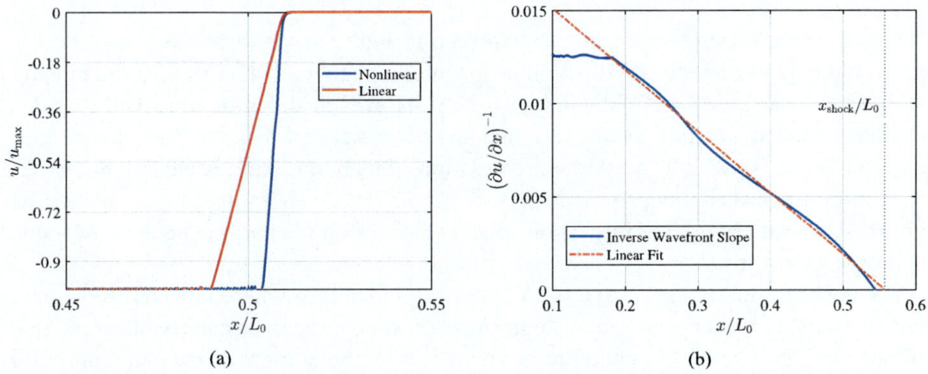

Fig. 1. (a) Nonlinear steepening of a rarefaction wave close to x_{shock} (b) Estimation of x_{shock} using the wavefront slope.

The same type of analysis is performed for the case of an inhomogeneous temperature distribution. A sinusoidal variation of 1 °C in temperature about a mean value of 370 °C is assumed and the corresponding values of Γ are computed using the same thermodynamic model. Figure 2(a) shows calculated propagation of the wave front into the medium (left axis) and the variation in Γ along the medium on the right. It can be seen that, unlike what can be noticed in Fig. 1(b), the slope of $(\partial u/\partial x)^{-1}$ is no longer constant as the wave continuously steepens and relaxes as the temperature varies along the medium. Remarkably, the rarefaction steepens to form a shock despite the thermodynamic states of medium display positive Γ in several locations. This observation can be justified by the fact that the bulk of the medium is in the nonlinear thermodynamic region, resulting in the overall effect that the wave steepens into a rarefaction shock wave. The effect of the local linear regions where $\Gamma \simeq 0$ is also visible by the inflection in $(\partial u/\partial x)^{-1}$ curve between the local extrema.

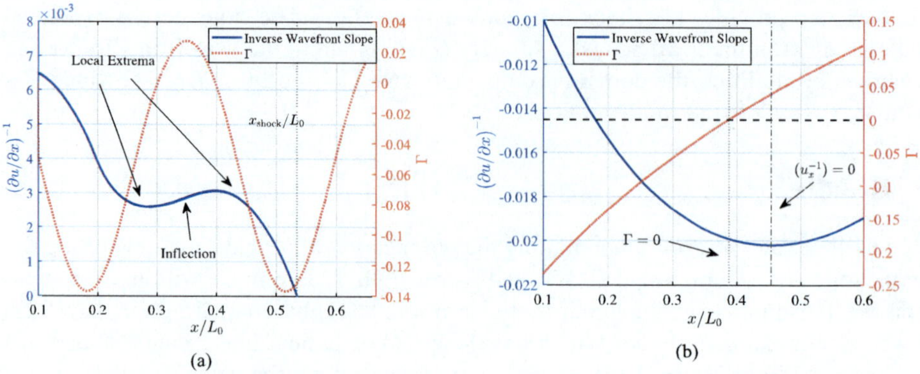

Fig. 2. (a) Steepening of the rarefaction wavefront in D6 with sinusoidally-varying temperature (b) Steepening of the rarefaction wavefront in D6 with linearly-varying temperature—Locations where $\Gamma = 0$ (······) and where $\left[(\partial u/\partial x)^{-1}\right]' = 0$ (- - -)

Another interesting feature can be observed in Fig. 2 at the locations where $\Gamma = 0$. One would expect that the wave would propagate unchanged if $\Gamma = 0$; thus the tangent of $(\partial u/\partial x)^{-1}$ at these locations must also be 0. However, it can be observed that this flattening of the wavefront always succeeds the point where Γ is zero. This can be seen clearly in Fig. 2(b), which shows the propagation of a compression wave in a fluid with temperature increasing linearly from 367.5 °C to 372 °C. Here, it is apparent that the compression initially relaxes into a non-ideal compression fan and steepens as the value of Γ becomes positive.

The vertical lines in Fig. 2(b) represent the axial locations where Γ is zero and where the wavefront starts steepening into a compression shock. It can be clearly observed that in a narrow region right after the point where $\Gamma = 0$, the compression wave continues to relax despite Γ becoming positive, indicating that the variation in the sound speed and density in this region has a significant influence on the wave distortion. Except in this region, the sign and magnitude of Γ remain the most important characteristics governing the nature of the nonlinear distortion of the wave.

4 Conclusions and Future Work

Preliminary results of a numerical investigation on the nonclassical propagation of nonlinear waves in case the medium is affected by temperature variations at the boundary of the domain indicate that, though the amplitude of the initial disturbance is crucial for nonlinear steepening, the minimum amplitude for shock formation is strongly dependent on the temperature distribution, and therefore, the variation of the fundamental derivative in the medium. Though Γ is the most important parameter influencing shock formation, it is observed that rarefaction shocks can steepen even in regions of positive nonlinearity before transitioning into a rarefaction fan. This study resulted in a tool that allows computing shock parameters such as formation distance and time for the case of a medium affected by varying temperatures. Further sensitivity studies will be performed to analyse the effect of temperature inhomogeneities on nonlinear wave propagation and shock formation. In addition, the applicability of this analysis to industrial problems will be explored.

References

1. Borisov, A.A., Borisov, A.A., Kutateladze, S., Nakoryakov, V.: Rarefaction shock wave near the critical liquid-vapour point. J. Fluid Mech. **126**, 59–73 (1983)
2. Colonna, P., Guardone, A., Nannan, N., Zamfirescu, C.: Design of the dense gas flexible asymmetric shock tube. J. Fluids Eng. **130**(3), 034501 (2008)
3. Colonna, P., Nannan, N., Guardone, A.: Multiparameter equations of state for siloxanes:[(ch3) 3-si-o1/2] 2-[o-si-(ch3) 2] i= 1,. . ., 3, and [o-si-(ch3) 2] 6. Fluid Phase Equilib. **263**(2), 115–130 (2008)
4. Fergason, S., Ho, T., Argrow, B., Emanuel, G.: Theory for producing a single-phase rarefaction shock wave in a shock tube. J. Fluid Mech. **445**, 37–54 (2001)
5. Haigh, A.A., Treeby, B.E., McCreath, E.C.: Ultrasound simulation on the cell broadband engine using the westervelt equation. In: International Conference on Algorithms and Architectures for Parallel Processing, pp. 241–252. Springer (2012)
6. Hamilton, M.F., Blackstock, D.T., et al.: Nonlinear acoustics, vol. 237. Academic Press, San Diego (1998)
7. Lauterborn, W., Kurz, T., Akhatov, I.: Nonlinear acoustics in fluids. In: Springer Handbook of Acoustics, pp. 265–314. Springer (2014)
8. Lemmon, E.W., Bell, I.H., Huber, M.L., McLinden, M.O.: NIST Standard Reference Database 23: Reference Fluid Thermodynamic and Transport Properties-REFPROP, Version 10.0, National Institute of Standards and Technology (2018)
9. Mathijssen, T., Gallo, M., Casati, E., Nannan, N., Zamfirescu, C., Guardone, A., Colonna, P.: The flexible asymmetric shock tube (FAST): a Ludwieg tube facility for wave propagation measurements in high-temperature vapours of organic fluids. Exp. Fluids **56**(10), 195 (2015)
10. Muralidharan, S., Sujith, R.: Shock formation in the presence of entropy gradients in fluids exhibiting mixed nonlinearity. Phys. Fluids **16**(11), 4121–4128 (2004)
11. Pinton, G.F., Dahl, J., Rosenzweig, S., Trahey, G.E.: A heterogeneous nonlinear attenuating full-wave model of ultrasound. IEEE Trans. Ultrason. Ferroelectr. Freq. Control **56**(3), 474–488 (2009)
12. Shevchenko, I., Kaltenbacher, B.: Absorbing boundary conditions for nonlinear acoustics: the westervelt equation. J. Comput. Phy. **302**, 200–221 (2015)

Similarity Parameters for Non-ideal One-Dimensional Isentropic Expansions

Camilla C. Conti(✉), Andrea Spinelli, and Alberto Guardone

Politecnico di Milano, Milan, Italy
{camillacecilia.conti,andrea.spinelli,alberto.guardone}@polimi.it

Abstract. This paper reports an investigation into similarity parameters for one-dimensional isentropic expansions of compressible flows in the non-ideal regime involving thermodynamic conditions in the close proximity of the vapor-liquid saturation curve and the critical point. Several expansions having different total pressure and temperature but sharing the same compressibility factor evaluated at total conditions were calculated using an Helmoltz energy-based fundamental relation of Span-Wagner type for several pure fluids of varying complexity and from different fluid classes.

The total compressibility factor was found to be a suitable similarity parameter for characterizing isentropic expansions of complex fluids in moderately high non-ideal conditions because it is associated with similar evolution of compressibility factor and fundamental derivative of gasdynamics along the process, thanks to the particular shape of their isolines in the temperature - entropy thermodynamic plane. Compressibility factor and fundamental derivative of gasdynamics can be seen as representative of information contained in thermal and caloric equations of state, respectively. Expansions sharing similar trends in these two parameters were found to be similar since similar volumetric and caloric behaviours rule the evolution of thermodynamic properties.

The use of the total compressibility factor as similarity parameter for isentropic expansions could allow to reduce the number of design parameters in preliminary sizing of thermodynamic cycles and components operating in the non-ideal compressible regime and could significantly decrease the experimental effort in wind tunnel calibration of probes employed in such flows.

Keywords: Similarity parameters · Non-ideal 1D isentropic expansions · Compressibility factor · Fundamental derivative of gasdynamics · Molecular complexity

1 Introduction

Non-ideal compressible fluid dynamics (NICFD) effects characterize fluid flows occurring close to the vapor-liquid saturation curve and the critical point. These

M. Pini et al. (Eds.): NICFD 2020, ERCOFTAC Series 28, pp. 26–35, 2021.
https://doi.org/10.1007/978-3-030-69306-0_4

were observed for flows of Siloxane MDM (octamethyltrisiloxane, $C_8H_{24}O_2Si_3$) in experimental campaigns reported in [1] and for Siloxane MM (hexamethyldisiloxane, $C_6H_{18}OSi_2$) in [2,3]. Pressure ratio P/P_T, where P is static pressure and P_T is total pressure, measured along supersonic expansions in converging-diverging nozzles was found to be dependant on stagnation conditions unlike ideal gases. Moreover, the total compressibility factor Z_T, defined in Eq. (1) where R is the gas constant, T_T is the total temperature and $\rho_T = \rho(T_T, P_T)$ is the stagnation density, was found to work well as a similarity parameter to univocally identify non-dimensional MM nozzle expansions in a given geometry for $Z_T \geq 0.6$.

$$Z_T = \frac{P_T}{RT_T\rho_T} \tag{1}$$

Pressure ratios of expansions having different total conditions but same total compressibility factor were experimentally measured to be within error bars of one another. For less ideal total conditions having $Z_T \leq 0.6$, the total compressibility factor alone was instead not sufficient to fully identify dimensionless expansions.

One-dimensional (1D) isentropic expansions were studied and documented here with the aim of further investigating the suitability of Z_T as a similarity parameter. This allowed to explore a wider region of the vapor phase than the one covered in [1–3], and to extend the analysis to fluids other than Siloxanes MDM and MM.

The identification of a similarity parameter in the context of non-ideal flows can provide precious information to reduce the number of variables for the preliminary design of thermodynamic cycles and components (e.g. turbomachinery) in various fields of application, such as organic Rankine cycle (ORC) power systems. Moreover, it can allow to drastically reduce the experimental effort in future wind tunnel calibration of pressure probes operating in the non-ideal compressible regime. Not all possible probe operating conditions would have to be reproduced, but only selected ones sharing the same similarity variable as real conditions would be required, entailing very significant time and costs savings.

The present paper is structured as follows. Section 2 presents the framework implemented for 1D isentropic expansions calculation. Section 3 illustrates results, in particular highlighting NICFD effects, assessing the performance of Z_T as a similarity parameter and evidencing the roles of compressibility factor and fundamental derivative of gasdynamics. Section 4 provides an interpretation for the effects of these two parameters on isentropic expansions from the point of view of thermal and caloric similarity. Section 5 draws the conclusions.

Table 1. Fluids selected for analysis with critical temperature T_C, critical pressure P_C, molar weight MW, number of atoms and chemical formula.

Type	Fluid	T_C [°C]	P_C [bar]	MW [kg/kmol]	N atoms	Chemical formula
Siloxanes	MM	245.55	19.31	162.38	27	$C_6H_{18}OSi_2$
	MDM	292.21	14.38	236.53	37	$C_8H_{24}O_2Si_3$
	D4	313.34	13.32	296.62	40	$C_8H_{24}O_4Si_4$
	D6	372.63	9.61	444.92	67	$C_{14}H_{42}O_5Si_6$
Alkanes	Butane	151.98	37.96	58.12	14	C_4H_{10}
	Pentane	196.55	33.70	72.15	17	C_5H_{12}
	Hexane	234.67	30.34	86.18	20	C_6H_{14}
	Octane	296.17	24.97	114.23	26	C_8H_{18}
	Decane	344.55	21.03	142.28	32	$C_{10}H_{22}$
Cyclic Alkanes	Cyclopentane	238.57	45.71	70.13	15	C_5H_{10}
	Cyclohexane	280.45	40.81	84.16	18	C_6H_{12}
Alkenes	Isobutene	144.94	40.10	56.11	14	C_4H_{10}
	Benzene	288.87	49.07	78.11	12	C_6H_6
Other Hydrocarbons	Toluene	318.60	41.26	92.14	15	C_7H_8
	Acetone	234.95	47.00	58.08	10	C_3H_6O
Halocarbons	R1234yf	94.70	33.82	114.04	9	$C_3H_2F_4$
	R218	71.87	26.40	188.02	11	C_3F_8
	R1233zd	165.60	35.73	130.50	9	$C_3H_2ClF_3$
	R227ea	101.75	29.25	170.03	11	C_3HF_7
	RC318	115.23	27.78	200.04	12	C_4F_8
Other Fluids	Water	373.95	220.64	18.02	3	H_2O
	Carbon Dioxide	30.98	73.77	44.01	3	CO_2
	Ammonia	132.25	113.33	17.03	4	NH_3

2 Calculation Framework

1D isentropic expansions were calculated for several total conditions sharing the same total compressibility factor Z_T. Expansions were determined independently of any particular geometry by letting the pressure ratio P/P_T vary in the range 0.99–0.05. Calculation was carried out for different levels of non-ideality, so for Z_T between 0.4 and 0.9, and for the pure fluids reported in Table 1. Several fluids from different families involving complex molecules (including Siloxanes, Alkanes, Cyclic Alkanes and Halocarbons) were considered. Simple molecules such as carbon dioxide, water and ammonia were also included in the analysis. Ten total conditions were considered for each Z_T. Total pressure and temperature (P_T, T_T) at same Z_T were chosen so as to cover the largest possible area in the vapor region of the $T - s$ thermodynamic plane of each fluid, where T and s are temperature and specific entropy. Figures 1a and 1b report chosen total conditions for exemplary fluids MM and CO_2 in their respective $T - s$ planes. An Helmholtz energy-based fundamental relation of Span-Wagner type implemented via the *FluidProp* thermodynamic library [4] was used to calculate all derived thermodynamic properties. The maximum considered pressure was

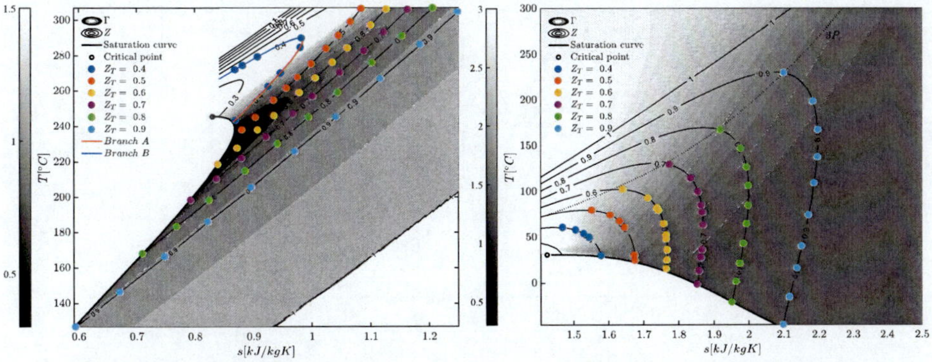

(a) $T-s$ diagram with Z and Γ contours for fluid MM.

(b) $T-s$ diagram with Z and Γ contours for CO_2.

Fig. 1. Total conditions selected for expansion calculations for fluids MM and CO_2 are shown by dots in the $T-s$ plane. Each color corresponds to a selected Z_T value, consistently with plots in Fig. 2.

three times the critical one and the maximum considered temperature was set to the upper validity limit of the thermodynamic model. Care was taken to ensure no entering in the two-phase region during any expansion. This is why, for some fluids, $Z_T = 0.4$ had to be excluded from calculation or expansions had to be interrupted at the saturation curve.

3 Results

3.1 Evidence of NICFD effects

1D calculations confirm the non-ideal dependence of the expansions on total conditions for all selected fluids. This is illustrated for fluids MM and CO_2 in Figs. 2a and 2b respectively, where the area ratio A^*/A along expansions is plotted as a function of P/P_T. Each curve is an expansion, evolving from $P/P_T = 1$ towards lower pressure ratios, with different colors corresponding to selected total compressibility values $Z_T = 0.4, 0.5, 0.6, 0.7, 0.8, 0.9$. As it can be seen, curves are not superposed, as would instead happen for an ideal gas behaviour.

Considering carbon dioxide, several expansions were found to at least partially proceed in the two-phase region. Thus, expansions plots were interrupted at the pressure ratio corresponding to two-phase region entry. In cases in which this occurred before the sonic condition, the whole expansion was instead removed from plots.

For brevity, results are shown only for exemplary fluids MM and CO_2, which were found to be well representative of high and low molecular complexity fluids, respectively. Indeed, qualitatively similar trends to MM were found for all investigated fluids except for those characterized by very low molecular complexity: water, carbon dioxide and ammonia. The behaviour of the latter three is instead well illustrated by the one of CO_2.

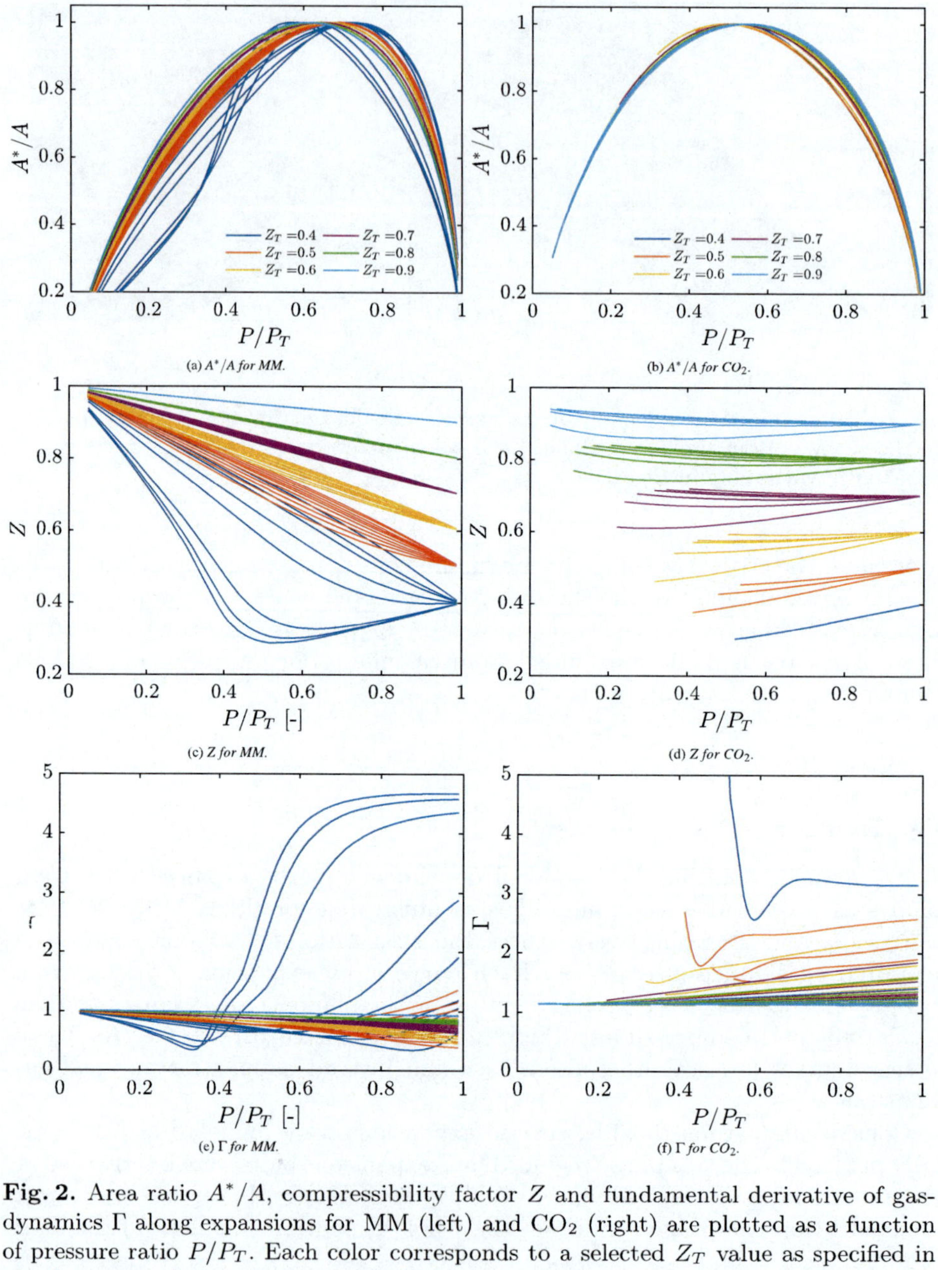

(a) A^*/A for MM. (b) A^*/A for CO_2. (c) Z for MM. (d) Z for CO_2. (e) Γ for MM. (f) Γ for CO_2.

Fig. 2. Area ratio A^*/A, compressibility factor Z and fundamental derivative of gas-dynamics Γ along expansions for MM (left) and CO_2 (right) are plotted as a function of pressure ratio P/P_T. Each color corresponds to a selected Z_T value as specified in plot legends and each line to a different total condition consistently with Fig. 1.

Referring to Fig. 2b, simple fluids show no particular pattern in expansions sharing the same total compressibility factor, which is therefore not a suitable similarity parameter to characterize the process. In case of complex fluids instead (Fig. 2a), A^*/A curves of expansions with the same Z_T are grouped together, except for strongly non-ideal conditions at $Z_T = 0.4$. In the latter case, although expansions are separated from other Z_T values, they are not superposed.

For a fixed geometry, which for 1D calculation corresponds to fixing the area ratio A^*/A, pressure ratio is larger for expansions occurring at more ideal conditions (lower total compressibility factor). Table 2 reports average pressure ratios $\overline{P/P_T}$ for the expansions at chosen Z_T for three selected area ratios: $A^*/A = 0.8$ - subsonic, $A^*/A = 1$ - sonic and $A^*/A = 0.7$ - supersonic. These correspond to Mach numbers for MM of ~ 0.6, 1 and ~ 1.5, respectively. As a quantitative example, if sonic conditions are considered, pressure ratio can vary from an average $\overline{P/P_T} = 0.68$ at $Z_T = 0.4$ to $\overline{P/P_T} = 0.61$ at $Z_T = 0.9$ for fluid MM, so by a non-negligible 11.5%. Table 2 also shows that the average pressure ratio has larger variation with the level of non-ideality when compressibility effects become more important, so towards A^*/A corresponding to higher Mach numbers.

3.2 Z_T as Similarity Parameter for Molecularly Complex Fluids

To quantify the performance of Z_T as a similarity parameter for complex fluids, the maximum percentage pressure ratio variation $\Delta^{Z_T}_{P/P_T}$ is calculated as:

$$\Delta^{Z_T}_{P/P_T} = \frac{\max\left[P/P_T(Z_T)\right] - \min\left[P/P_T(Z_T)\right]}{\min\left[P/P_T(Z_T)\right]} \cdot 100 \qquad (2)$$

where $\max\left[P/P_T(Z_T)\right]$ and $\min\left[P/P_T(Z_T)\right]$ refer to maximum and minimum values of pressure ratio registered at a specified A^*/A value among all expansions sharing the same Z_T. It is therefore an indication of the spread in pressure ratio values in a fixed geometry for expansions characterized by the same total compressibility factor but different (P_T, T_T). Considering Siloxane MM, Table 2 and Fig. 3 report $\Delta^{Z_T}_{P/P_T}$ values at chosen Z_T for the three selected area ratios previously considered. At sonic conditions ($A^*/A = 1$) and $Z_T = 0.4$, $\Delta^{Z_T}_{P/P_T} = 28.52\%$. This is a very large value consistent with previous qualitative observations. As expected, there are significant improvements as more ideal conditions at higher total compressibility factor are approached, with $\Delta^{Z_T}_{P/P_T} = 0.57\%$ at $Z_T = 0.6$ and a further decrease to $\Delta^{Z_T}_{P/P_T} = 0.31\%$ at $Z_T = 0.9$. Considering other area ratios and the other investigated complex fluids, the same trend in $\Delta^{Z_T}_{P/P_T}$ with Z_T is evidenced. The percentage pressure ratio variation increases as the Mach number increases at same Z_T also due to the lower value of the pressure ratio at the denominator (Eq. (2)).

Siloxane MDM, a heavier fluid with higher complexity than MM, in its same family and with similar molecular structure, has overall lower values, with $\Delta^{Z_T}_{P/P_T} = 0.28\%$ at $Z_T = 0.6$ and $\Delta^{Z_T}_{P/P_T} = 0.15\%$ at $Z_T = 0.9$ for $A^*/A = 1$. Lower values of $\Delta^{Z_T}_{P/P_T}$ were also found for Siloxane D6 with respect to the simpler D4. Analogously, considering the Alkanes family, Z_T performance was found to improve as molecular mass and complexity increase leading to lower $\Delta^{Z_T}_{P/P_T}$ values, except at strongly non-ideal total conditions with $Z_T = 0.4$. In this case instead, total compressibility factor was found to perform worse for more complex fluids.

Table 2. Average pressure ratio $\overline{P/P_T}$ and percentage difference in pressure ratio $\Delta^{Z_T}_{P/P_T}$ for fluid MM at different Z_T values and selected area ratios $A^*/A = 0.8$ - subsonic, $A^*/A = 1$ - sonic and $A^*/A = 0.7$ - supersonic.

MM	$Z_T = 0.4$		$Z_T = 0.6$		$Z_T = 0.9$	
	$\overline{P/P_T}$	$\Delta^{Z_T}_{P/P_T}$	$\overline{P/P_T}$	$\Delta^{Z_T}_{P/P_T}$	$\overline{P/P_T}$	$\Delta^{Z_T}_{P/P_T}$
$A^*/A = 0.8$, **sub**	0.87	15.55	0.88	0.32	0.85	0.00
$A^*/A = 1$, **sonic**	0.68	28.52	0.66	0.57	0.61	0.31
$A^*/A = 0.7$, **sup**	0.37	35.02	0.30	3.88	0.26	0.73

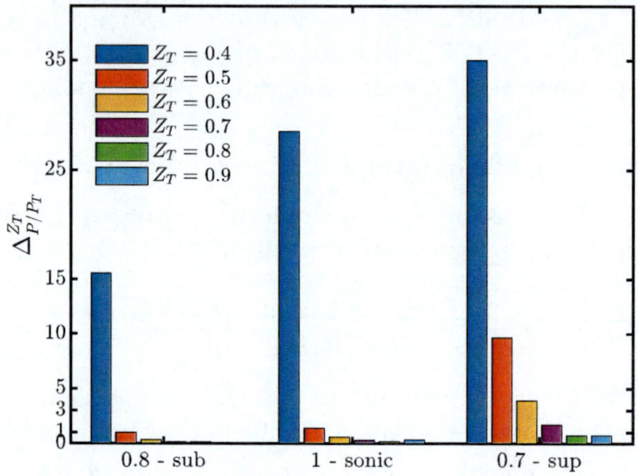

Fig. 3. Percentage difference in pressure ratio $\Delta^{Z_T}_{P/P_T}$ for fluid MM at different Z_T values and selected area ratios $A^*/A = 0.8$ - subsonic, $A^*/A = 1$ - sonic and $A^*/A = 0.7$ - supersonic.

It should be pointed out that within other classes of complex fluids evaluated in the present analysis, this improvement with molecular complexity in the performance of Z_T as a similarity parameter was not as evident as for Siloxanes or Alkanes. This is because molecular structure can change significantly within the same fluid class (consider Halocarbons, for example), with important impacts on thermodynamic properties that make any analysis based only on molecular complexity misleading.

3.3 Z and Γ along Expansions

Figure 2 also reports plots of compressibility factor Z and fundamental derivative of gasdynamics Γ (defined in Eq. (3) where c is speed of sound and v is specific volume) as a function of pressure ratio P/P_T for both MM and CO_2.

$$\Gamma = 1 + \frac{c}{v}\left[\left(\frac{\partial c}{\partial P}\right)_s\right] \tag{3}$$

It is remarkable that, considering a complex fluid, total conditions having same Z_T also have similar Z and Γ trends and values along the expansion, as seen in Figs. 2c and 2e. As less ideal conditions at lower Z_T are approached, however, larger differences in Γ and/or Z values are found. This is linked to the particular shape of compressibility factor and fundamental derivative isolines in the $T-s$ thermodynamic plane for complex fluids. Far away from the critical point, iso-Z are lines parallel to one another, as seen in the $T-s$ diagram in Fig. 1a. Therefore, if expansions with different total conditions share the same Z_T, they also share close values of Z all along the process, since expansions are simply vertical lines in the $T-s$ plane. Moreover, iso-Γ curves are also parallel away from the critical point - not only between themselves, but also parallel to iso-Z curves. As a consequence, if expansions share the same Z_T, they not only share similar Z, but also similar Γ values along the process.

At less ideal conditions, both Z and Γ isolines are no longer simple straight lines, but they form an enclosed region. Starting from the saturation curve and moving along an isoline into the vapor region, the path is initially a straight line towards higher temperature and entropy. This is exemplified as *Branch A* in Fig. 1a on the $Z = 0.4$ isoline. A maximum entropy point is reached and the path along the isoline then bends back towards the saturation curve and lower entropy and temperature values (*Branch B* in Fig. 1a). Iso-Γ curves show this enclosed region as the critical point is approached from low temperatures. Iso-Z curves instead show this behaviour close to the critical point aswell, but at higher pressures than iso-Γs. If expansions start on a point on *Branch B* of the iso-Γ or iso-Z line, then the trend in Γ or Z along the expansion will be non-monotone. Trends in Z and Γ are thus no longer similar to other expansions at same Z_T starting on *Branch A*. This can be seen in Figs. 2c and 2e for $Z_T = 0.4$.

As previously mentioned, if simple molecules are considered instead, pressure ratio trends are not successfully characterized by the value of Z_T. Interestingly, if Z_T is the same, the compressibility factor trend along expansions is also comparable, whilst Γ does not show any common pattern.

Again, this is linked to the particular shape of the isolines in the $T-s$ thermodynamic plane, which is significantly different with respect to complex fluids. In the $T-s$ diagram in Fig. 1b, iso-Z curves are still fairly parallel to one another along their *Branch A*, which is almost a vertical line. As a consequence, the value of Z does not vary much along an expansion, differently to what happens for complex fluids. This also why expansions with same Z_T have similar Z trends too. Considering iso-Γ curves instead, they are not at all parallel to iso-Zs, contrarily to complex fluids. This is responsible for the fact that total conditions sharing the same Z_T do not have similar values and trends of Γ if simple molecules are considered.

The above observations indicate that expansion similarity is linked to similar trends and values in both the compressibility factor and the fundamental derivative. This occurs for complex fluids at moderately high levels of non-ideality for expansions sharing the same Z_T thanks to the particular shape of Z and Γ isolines. Thus, Z_T is here a good similarity parameter because it entails similar values and trends in Z and Γ aswell. At strongly non-ideal conditions or when simple fluids are considered instead, having the same Z_T does not lead to similar Z and Γ along expansions and it is not a satisfactory similarity parameter.

A theoretical interpretation for the effect of compressibility factor and fundamental derivative on isentropic expansions is provided in the following section.

4 Thermal and Caloric Behaviour Similarity

For a complete thermodynamic description of a single constituent fluid, either a fundamental relation or a thermal and a caloric equation of state need to be provided. The thermal equation is a description of the volumetric behaviour of the fluid, whilst the caloric equation accounts for the way in which energy is stored and released by the fluid as it undergoes a process.

In the search for similarity parameters characterizing non-ideal isentropic expansions where thermodynamic behaviour is clearly important, it is intuitive to look for quantities that are able to represent the information given by the thermal equation of state and the caloric one. If these parameters were to be the same along different expansions, then it would be reasonable to assume that those expansions are, infact, similar, since similar volumetric and caloric behaviours rule the evolution of thermodynamic properties.

The compressibility factor Z can be considered an indication of the volumetric behaviour of a fluid with respect to an ideal gas at same pressure and temperature and is representative of the thermal equation of state. Thus, if two expansions share the same compressibility factor, both at total conditions and during the expansion process, then they have similar volumetric behaviour throughout.

Considering the caloric equation of state, the fundamental derivative can be considered as a suitable similarity parameter in the case of isentropic expansions. Indeed, it is a quantitative measure of the variation of the speed of sound in isentropic transformations, so it is intrinsically defined to characterize the type of processes here analyzed. Moreover, it is explicitly dependent on molecular complexity [5], so it accounts for the number of degrees of freedom in which a molecule can store energy during the expansion. This is precisely the information held in the caloric equation of state and if expansions share similar trends and values in Γ, then they can be expected to depend on similar caloric behaviour.

5 Conclusions

1D isentropic expansions for non-ideal flows of several different fluids with varying degree of molecular complexity were investigated and reported in the frame of the present paper. Findings indicate that, for complex fluids in moderately high non-ideal conditions, expansions are similar if total conditions share the same total compressibility factor. This is because trends and values of compressibility factor and fundamental derivative of gasdynamics are also similar along the process thanks to the particular shape of their isolines in the $T-s$ thermodynamic plane. These parameters are representative of information stored in thermal and caloric equations of state, suggesting that expansions are similar because they share similar volumetric and caloric behaviour throughout the process.

References

1. Spinelli, A., et al.: Experimental evidence of non-ideal compressible effects in expanding flow of a high molecular complexity vapor. Exp. Fluids **59**, 126 (2018)
2. Cammi, G., Conti, C.C., Spinelli, A., Guardone, A.: Experimental characterization of nozzle flow expansions of siloxane mm for orc turbines applications. In: 5th International Seminar on ORC Power Systems, Athens, Greece (2019)
3. Cammi, G., Conti, C.C., Spinelli, A., Guardone, A.: Experimental characterization of nozzle flow expansions of siloxane MM for ORC turbines applications. Energy (2020). https://doi.org/10.1016/j.energy.2020.119249
4. Colonna, P., Van der Stelt, T.P., Guardone, A.: FluidProp: a program for the estimation of thermo physical properties of fluids (2004)
5. Harinck, J., Guardone, A., Colonna, P.: The influence of molecular complexity on expanding flows of ideal and dense gases. Phys. Fluids **21**, 086101 (2009)

Non-ideal Effects in Compressible Swirling Flows

Francesco Tosto, Andrea Giuffre', Piero Colonna, and Matteo Pini(✉)

Propulsion and Power, Delft University of Technology, Delft, The Netherlands
m.pini@tudelft.nl

Abstract. In this work we examine the behavior of non-ideal compressible swirling flows. Based on a first-principle analysis, we derive a generalized expression of the corrected flow per unit area as function of the isentropic exponent, characteristic Mach numbers, and swirl parameter. The calculated trends of the corrected flow with respect to these parameters, validated against results from high-fidelity computations, are used to infer physical insights on the behavior of swirling flows in turbomachinery cascades. The results suggest that fluid flows characterized by low values of the isentropic exponent show swirling behaviors that are substantially different than those exhibited by perfect gases. Ultimately, this can make the design of efficient turbomachines operating close to the critical point particularly challenging.

Keywords: Isentropic exponent · Swirling flow · Corrected flow function

1 Introduction

Swirling flows, namely flows characterized by a non-negligible circumferential velocity component, are ubiquitous in internal flow devices. Swirling in turbomachines is typically induced by the circumferential motion the flow is subject to when passing through a blade row or a radial vaneless duct, e.g. a radial diffuser. The combination of swirl and compressibility in turbomachinery leads to flow features beyond those encountered in the examination of either phenomenon separately and that can have large impact on their performance. The key parameter governing the behavior of a turbomachinery stage is the well-known corrected flow per unit area, which is a direct measure of its flow capacity [3]. This quantity is used to determine, among others, the response of a swirling flow to changes in passage area and is thus highly suited to characterize the actual behavior of a turbomachine stage, as for instance its choking condition. As outlined in recent works [2,7] compressibility effects in turbomachinery blade rows operating in the non-ideal compressible flow regime can be greatly enhanced or mitigated depending on whether the so-called generalized isentropic exponent [5] is higher or lower than the heat capacity ratio of the fluid in the limit of dilute gas. The occurrence of non-ideal gas effects in swirling flows is then expected to have large impact on the choking condition of a turbomachinery stage, thus on the value of the corrected flow per unit area. This paper presents a fundamental investigation aimed at examining,

M. Pini et al. (Eds.): NICFD 2020, ERCOFTAC Series 28, pp. 36–43, 2021.
https://doi.org/10.1007/978-3-030-69306-0_5

from a qualitative standpoint, characteristic features of isentropic non-ideal compressible swirling flows typically encountered in turbomachines with the ultimate goal of revealing potential opportunities for their optimal design.

2 Background

Figure 1 shows the schematic of a typical swirling flow configuration and the nomenclature used throughout the paper. The canonical expression of the corrected flow per unit area for ideal gas flows [3] can be rearranged to highlight the role of the swirl velocity component. The canonical equation reads

$$\dot{m}_{\mathrm{corr}} = \frac{\dot{m}\sqrt{RT_{\mathrm{t}}}}{p_{\mathrm{t}}A\sqrt{\gamma}} = \sqrt{\frac{T}{T_{\mathrm{t}}}\frac{\rho}{\rho_{\mathrm{t}}}}\frac{u_m}{\sqrt{\gamma RT_{\mathrm{t}}}} = M_{\mathrm{m}}f(M), \tag{1}$$

where M_{m} is the meridional Mach number. By introducing the generalized isentropic exponent, defined as

$$\gamma_{pv} = -\frac{v}{p}\left(\frac{\partial p}{\partial v}\right)_s = -\frac{v}{P}\frac{c_p}{c_v}\left(\frac{\partial P}{\partial v}\right)_T, \tag{2}$$

and related to the fundamental derivative of gas dynamics Γ [8] through the relation [4]

$$\Gamma = \frac{1}{2}\left[\gamma_{pv} + 1 - \frac{v}{\gamma_{pv}}\left(\frac{\partial \gamma_{pv}}{\partial v}\right)_s\right], \tag{3}$$

one can rewrite (1) as function of γ_{pv} or Γ. When the fluid state is sufficiently far from the critical point, the partial derivative in Eq. 3 can be neglected and a linear trend of γ_{pv} with Γ is established. Such approximation has been exploited, for example, in the studies documented in [2,9].

By using the same approximation, Eq. (1) can therefore be casted in a form that is suited to point out the effect of flow non-ideality. By resorting to the generalized isentropic exponent γ_{pv} [2], this yields

$$\dot{m}_{\mathrm{corr}} = \frac{\dot{m}\sqrt{Z_{\mathrm{t}}RT_{\mathrm{t}}}}{p_{\mathrm{t}}A\sqrt{\gamma_{pv}}} = \frac{M_m}{\left(1+\frac{\gamma_{pv}-1}{2}M^2\right)^{\frac{\gamma_{pv}+1}{2(\gamma_{pv}-1)}}}, \tag{4}$$

where $Z_{\mathrm{t}} = p_{\mathrm{t}}/(\rho_{\mathrm{t}}RT_{\mathrm{t}})$ is the flow compressibility factor defined at total conditions. Eventually, after further algebraic manipulation

$$\dot{m}_{\mathrm{corr}} \simeq \frac{\dot{m}}{\rho_{\mathrm{t}}c_{\mathrm{t}}A} = \left(1+\frac{\gamma_{pv}-1}{2}M^2\right)\sqrt{\frac{M^2}{1+\frac{\gamma_{pv}-1}{2}M^2} - \left(\frac{u_\theta}{\sqrt{\gamma_{pv,\mathrm{t}}Z_{\mathrm{t}}RT_{\mathrm{t}}}}\right)^2}, \tag{5}$$

where the second term under the square root is referred to as swirling parameter and denotes the amount of swirl of a given flow. Equation 5 clearly shows the relation

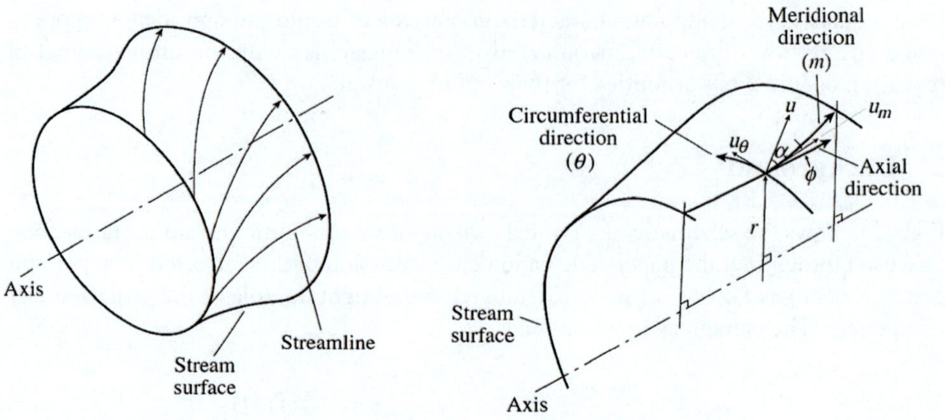

Fig. 1. Axisymmetric swirling flow in a duct with variable cross section and coordinate system [3].

between the corrected flow per unit area and the swirling parameter. Using the isentropic relation $p_t/p = f(M, \gamma_{pv})$, Eq. 4 can also be rewritten in terms of pressure ratio and flow (swirl) angle, namely

$$\dot{m}_{\text{corr}} = \left(\frac{p_t}{p}\right)^{\frac{\gamma_{pv}+1}{2\gamma_{pv}}} \sqrt{\frac{2}{\gamma_{pv}-1}\left[\left(\frac{p}{p_t}\right)^{\frac{1-\gamma_{pv}}{\gamma_{pv}}}-1\right]} \cos\alpha. \tag{6}$$

Note that the accuracy of the results obtained by applying the above equations depends on the actual variation of γ_{pv} along an expansion process and is therefore related to the extent of flow non-ideality, namely on the relative difference between γ and γ_{pv}, and to the flow Mach number.

3 Technical Approach

Equations (5) and (6) constitute the analytical models used to investigate the impact of non-ideal flow effects on the behavior of swirling flows. The study is performed by considering expansion processes occurring in an exemplary turbomachine operating with a paradigmatic fluid that can exhibit substantial variability of degree of non-ideality, herein evaluated as the deviation of γ_{pv} from γ of the fluid in the dilute gas state. The contours of γ_{pv} for siloxane MM and CO_2 are shown in Fig. 2 to illustrate the range of variation of the isentropic exponent for hypothetical expansion processes occurring in the non-ideal compressible flow regime. For instance, pre-expansion processes at the impeller inlet of supercritical CO_2 compressors usually occur in thermodynamic regions where $\gamma_{pv} > \gamma$, while flow expansions in an ORC turbine operating with MM can entail values of γ_{pv} lower or higher than γ. Table 1 lists the main fluid properties of the two fluids under investigation along with the lower and upper values of γ_{pv} occurring in the thermodynamic region bounded by $p < 2.5p_{cr}$ and $s > s_{cr}$. Note that the higher the

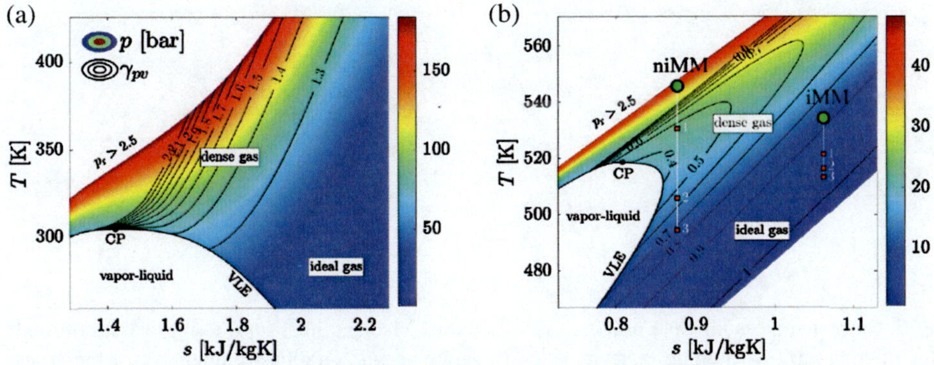

Fig. 2. Thermodynamic temperature-entropy diagrams of CO_2 (a) and siloxane MM (b). Pressure and γ_{pv} contours are reported. CP and VLE denote critical point and saturation vapor-liquid curve. Isentropic processes iMM and niMM discussed in Sect. 4.2 are reported in figure (b).

Table 1. Characteristic fluid properties of air, carbon dioxide and siloxane MM. γ_∞ is the specific heat ratio calculated in the dilute gas state. The lower and upper bound of γ_{pv} are those occurring in the thermodynamic region defined by $s > 1.01 s_{\mathrm{cr}}$ and by the thermal stability limits of each molecule.

	M_{mol} [kg/kmol]	T_{cr} [K]	p_{cr} [bar]	γ_∞	$\gamma_{pv,\mathrm{min}}$	$\gamma_{pv,\mathrm{max}}$
Air	28.96	132.82	38.50	1.40	1.44	2.85
CO2	44.01	304.13	73.77	1.29	0.86	4.33
MM	162.3	518.70	19.31	1.03	0.39	4.75

molecular complexity of the fluid, the lower the γ_{pv} value that the fluid exhibits in the considered thermodynamic region. Furthermore, fluids made by complex molecules can feature a non monotonic trend of γ_{pv} along an isentropic process, as shown in Fig. 2b.

The trends of the corrected flow per unit area are computed as a function of M_{m} or pressure ratio p/p_{t} and for different values of total Mach number M, swirl angle α, and normalized swirl velocity component $\hat{u}_\theta = u_\theta / \sqrt{\gamma_{pv,\mathrm{t}} Z_{\mathrm{t}} R T_{\mathrm{t}}}$.

4 Results

4.1 Physical Insights

Figure 3 shows the trend of corrected flow per unit area as a function of the meridional Mach number for three values of generalized isentropic exponent γ_{pv}. In the charts, lines of constant total Mach number, flow angle, and swirl parameter are also displayed. Several physical insights relevant to turbomachinery applications can be derived from inspection of these charts.

First, it can be seen that the vertical line corresponding to $M_m = 1$ passes through the maxima of the curves at constant $\hat{u}_\theta$. This means that choking in swirling flows is actually reached when the magnitude of the meridional velocity component equals that

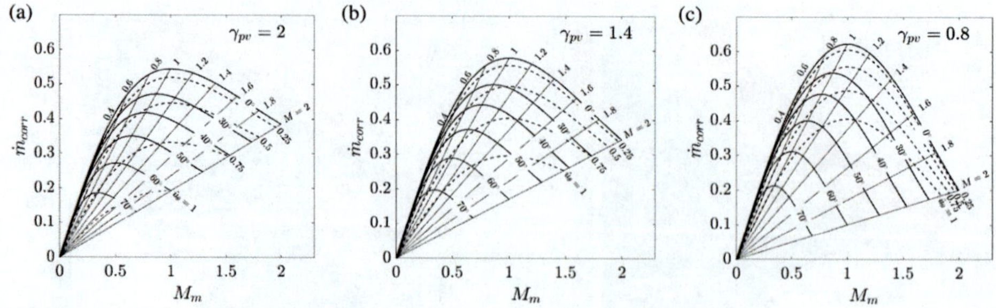

Fig. 3. Corrected mass flow per unit area vs meridional Mach number for different values of total Mach number M, swirl angle α, swirl velocity parameter $\hat{u}_\theta$. Graphs are reported for the cases (a) $\gamma_{pv} = 2$, (b) $\gamma_{pv} = 1.4$ and (c) $\gamma_{pv} = 0.8$.

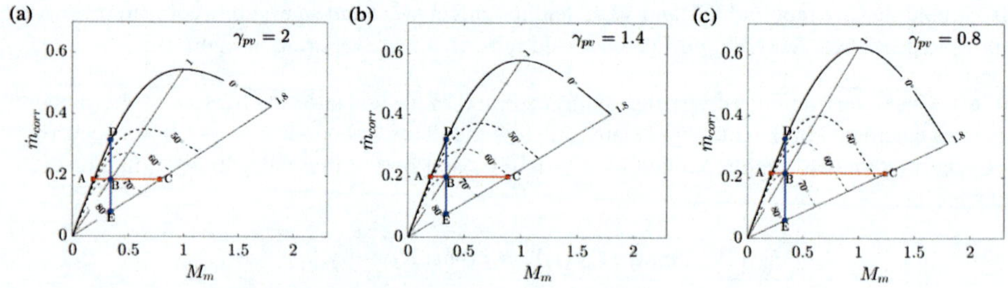

Fig. 4. Characteristic flow state trajectories on the $\dot{m}_{\rm corr} - M_{\rm m}$ diagrams. Line A→C illustrates a process at constant $\dot{m}_{\rm corr}$; line D→E illustrates a process at constant $M_{\rm m}$. Graphs are reported for the cases (a) $\gamma_{pv} = 2$, (b) $\gamma_{pv} = 1.4$ and (c) $\gamma_{pv} = 0.8$.

of the speed of sound and the flow is already in the supersonic regime, i.e. $M > 1$. The value of total Mach number at which choking occurs strongly depends on the γ_{pv} value. For instance, by assuming $\hat{u}_\theta = 0.75$, choking is achieved at $M \approx 1.2$ for $\gamma_{pv} = 0.8$ and at higher Mach numbers for larger values of γ_{pv}. This finding has important consequences for turbomachinery design. Given that the line at $M_m = 1$ represents the physical limit of operation of a turbomachine vane providing swirl to the flow ($M_m > 1$ would entail the impossibility of the flow to adapt to outlet conditions), it is common practice to design swirling devices in the transonic/supersonic regime with sufficient margin from that limit to avoid controllability issues. If the vane operates in a thermodynamic region whereby $\gamma_{pv} < \gamma$ the condition $M_m = 1$ would be reached sooner, meaning that the component cannot be designed for high total Mach numbers. Conversely, vanes designed to operate in regions whereby, on average, $\gamma_{pv} > \gamma$ can handle highly supersonic flow. However, a quantitative estimate of suitable ranges of Mach number for safe operation could only be estimated once the amount of swirl is known.

The last considerations have important implications in terms of flow behavior within actual turbine cascades. The insights can be better appreciated by examination of Fig. 4, that shows only a subset of the curves previously displayed. The lines A-B-C and D-E identify two different gas-dynamic processes. Notably, the horizontal line A-B-C at

Table 2. Boundary conditions used in the CFD simulations.

	$p_{\mathrm{t,in}}$ [bar]	$T_{\mathrm{t,in}}$ [K]	$p_{\mathrm{out},1}$ [bar]	$p_{\mathrm{out},2}$ [bar]	$p_{\mathrm{out},3}$ [bar]
iMM	9.66	534.26	4.13	3.44	2.80
niMM	40.00	545.67	22.48	14.45	10.23

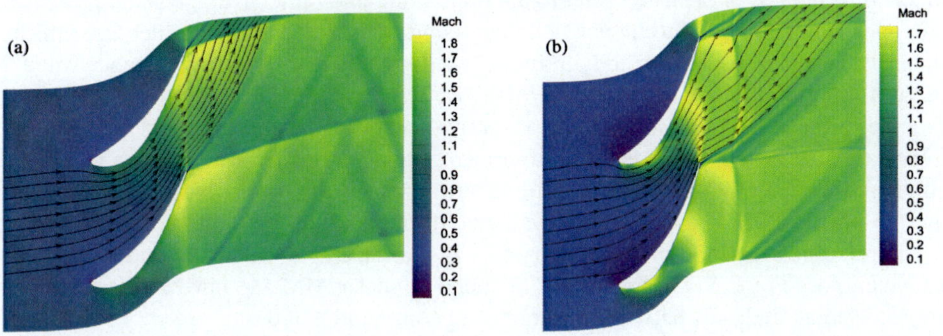

Fig. 5. Mach number contours and streamlines computed by 3D RANS at midspan for (a) iMM, (b) niMM close to chocking (case 3).

constant corrected flow is approximately representative of the initial (point B) and final states of a post-expansion (point C) or post-compression (point A) process occurring downstream of a choked turbine row. Following the line B-C, it can be observed that the flow deflects towards the meridional direction while expanding from choking conditions to $M = 1.8$. The magnitude of the deflection amounts to more than 20° when $\gamma_{pv} = 0.8$ and less than 10° when $\gamma_{pv} = 2.0$. In other words, the flow post-expanding downstream of a turbine row operating in the proximity of the critical point, where $\gamma_{pv} < \gamma$, is subject to deviations that are substantially higher than those that would occur if the same flow behaved as an ideal gas. Because of that, the assumption of small or negligible flow deviation, underlying most of the meanline design methods, is essentially incorrect in non-ideal flow conditions and a careful evaluation of this quantity becomes necessary during early stage of turbomachinery design.

The vertical line D-E exemplifies a flow expansion with the total Mach number increasing from nearly 0.5 to 1.8 at constant $M_{\mathrm{m}} = 0.3$, a behavior qualitatively representative of the flow process in a supersonic axial turbine vane designed at constant flow coefficient and with negligible post-expansion. To achieve this acceleration, the swirl angle rises from 0° to roughly 80° in all cases and the corrected flow reduces because of the corresponding increase of passage area. The reduction of corrected flow and the associated increase of annulus area is more significant as γ_{pv} decreases, as a direct consequence of the increase in volumetric flow ratio ρ_{t}/ρ. This finding corroborates the previous results in that the design of an efficient turbine row expanding a flow in near-critical thermodynamic conditions is arguably much more challenging to accomplish.

4.2 Verification

Results obtained by means of high-fidelity CFD [1] are utilized to assess the accuracy of the previous trends. 3D RANS computations of the flow past converging turbine vanes designed to operate with siloxane MM in ideal (case iMM) and non-ideal (case niMM) compressible flow regime are performed. The boundary conditions, specified in terms of inlet total pressure, inlet total temperature and outlet average static pressure are reported in Table 2. Turbulence boundary conditions, set in terms of inlet turbulence intensity, eddy viscosity ratio and turbulent Prandtl number, are fixed to reference values for all the simulations: $k = 5\%, \mu/\mu_t = 10, Pr_t = 1$. Total variation diminishing (TVD) schemes are adopted to discretize the advective and turbulent fluxes, while the viscous fluxes are discretized with a central difference scheme. $k - \omega$ SST turbulence model is employed together with adequate cell clustering near walls to guarantee $y^+ < 1$. The thermo-physical fluid properties are evaluated through a look-up table method coupled to the reference thermodynamic library [6]. A grid size of $5 \cdot 10^6$ cells and a look-up table composed by $2.5 \cdot 10^6$ and $4 \cdot 10^6$ elements for case iMM and niMM, respectively, are selected as trade-off between accuracy and computational cost.

The corrected mass flow per unit area is evaluated at midspan radius in three different sections, namely at the inlet, throat and outlet boundaries. The expansion ratio across the blade rows is progressively increased from the design point (case 1) to investigate post-expansion effects (cases 2–3). The average values of γ_{pv} and local flow deflection α computed by CFD at the outlet section are used to evaluate $\dot{m}_{\text{corr}}$ by means of Eq. (6). Figure 6 shows the comparison of the results obtained with the two models, together with the variation of the generalized isentropic exponent along the corresponding isentropic processes depicted in Fig. 2b. Several considerations can be drawn by examination of these charts.

First, provided that there is sufficient margin from the actual chocking limit of the cascade, i.e. up to point 2 for the case niMM, where $M_m \approx 0.9$, predictions made by the analytical models are qualitatively in line with the CFD results. Moreover, CFD trends are in agreement with the analytical results in that a lower variation of $\dot{m}_{\text{corr}}$ from inlet to outlet can be expected when increasing γ_{pv}.

The CFD simulations confirm the validity of the analysis regarding the post-expansion process reported in the previous paragraph. In this regard, the use of converging-diverging blade channels in turbine stages where the target expansion entails an average value of $\gamma_{pv} < \gamma$ can be advantageous to better control the flow deviation rather than to improve the fluid-dynamic efficiency.

Lastly, the ideal expansion process represented in Fig. 4 by the D-B-C line can be compared to the actual expansion in a converging turbine vane designed at constant flow coefficient and operating at off-design, i.e. the points inlet-throat-outlet (2–3) in Fig. 6. One can notice that in the actual expansion process the sonic state is systematically shifted rightwards, since chocking occurs at a lower flow angle than that of point B. In view of this observation, point B can be deemed to reproduce more closely the outlet state at design conditions, i.e. when the post-expansion is negligible and the total Mach number is close to one.

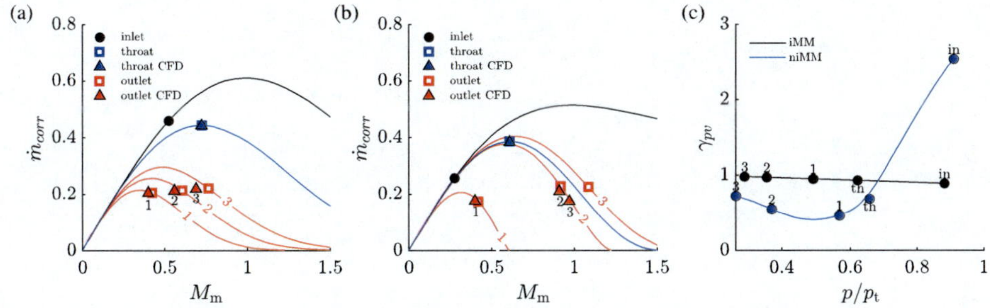

Fig. 6. Validation of test cases (a) iMM and (b) niMM with CFD. $\dot{m}_{\text{corr}}(M_{\text{m}})$ curves in red are computed with values of local flow deflection α and γ_{pv} at the stator outlet obtained from CFD. The black $\dot{m}_{\text{corr}}(M_{\text{m}})$ curve refers to the stator inlet conditions. (c) Variation of γ_{pv} with pressure drop along the isentropic processes iMM and niMM.

5 Conclusions

Non-ideal compressible swirling flows can show behaviors that are substantially different than those exhibited by perfect gases. The results of the work pointed out that, for fluids made by complex molecules, post-expansion effects in turbomachine vanes are greatly enhanced and choking can occur sooner if the expansion process occurs in the proximity of the critical point, where γ_{pv} assume values lower or significantly lower than γ_∞. Further analyses are needed to gain additional physical insights and to determine the relative implications for the optimal design of turbomachinery operating with non-ideal compressible flows.

References

1. ANSYS: Ansys workbench, release 19.3, ANSYS, canonsburg, pa (2019)
2. Baumgärtner, D., Otter, J.J., Wheeler, A.P.S.: The effect of isentropic exponent on transonic turbine performance. J. Turbomach. **142**(8), 081007/1–081007/10 (2020)
3. Greitzer, E.M., Tan, C.S., Graf, M.B.: Internal Flow: Concepts and Applications. Cambridge University Press, Cambridge (2004)
4. Henderson, L.R.F.: General laws for propagation of shock waves through matter. In: Handbook of Shock Waves, pp. 143–183. Elsevier (2001)
5. Kouremenos, D.A., Antonopoulos, K.A.: Isentropic exponents of real gases and application for the air at temperatures from 150k to 450k. Acta Mech. **65**(1), 81–99 (1987)
6. Lemmon, E.W., Bell, I.H., Huber, M.L., McLinden, M.O.: NIST standard reference database 23: Reference fluid thermodynamic and transport properties-REFPROP, version 10.0, National Institute of Standards and Technology (2018)
7. Romei, A., Vimercati, D., Persico, G., Guardone, A.: Non-ideal compressible flows in supersonic turbine cascades. J. Fluid Mech. **882**, A12 (2020)
8. Thompson, P.A.: A fundamental derivative in gasdynamics. Phys. Fluids **14**(9), 1843 (1971)
9. Wheeler, A.P.S., Ong, J.: The role of dense gas dynamics on organic rankine cycle turbine performance. J. Eng. Gas Turbines Power **135**(10), 102603/1–102603/9 (2013)

Supercritical Combustion

Generic Numerical Test Case to Understand Cryogenic Methane Combustion Dynamics

Stefan Fechter(✉), Tim Horchler, Sebastian Karl, and Klaus Hannemann

Spacecraft Department, German Aerospace Center, Institute of Aerodynamics and Flow Technology, Göttingen, Germany
{stefan.fechter,tim.horchler,sebastian.karl,klaus.hannemann}@dlr.de

Abstract. The accurate numerical simulation of high-pressure rocket combustion chambers is a key element for the design of future space transportation vehicles. In recent years investigations focus more and more on methane-based combustion chambers for which many thermodynamic and combustion issues are still unknown. Experimental investigations indicate that we might have to deal with a close interaction of non-ideal equation-of-state effects together with complex kinetic schemes.

On the numerical side the main challenges include the consistent approximation of thermodynamic effects within a broad temperature range from cryogenic injection up to high temperature effects at combustion. Nowadays simple numerical test cases are missing to study and understand the basic effects during the injection and combustion process, especially with methane as propellant. This open question is addressed by the definition of a simple test case for cryogenic methane combustion and its analysis by applying a range of thermodynamic and kinetic models.

Keywords: Rocket combustion · Methane · Non-ideal EOS · Non-ideal mixture effects · Cubic EOS

1 Introduction

The accurate numerical simulation of jet atomization and high-pressure combustion effects at cryogenic temperatures in rocket combustion chambers is a key element for the design of future space transportation vehicles. In the last years the focus of the research was on the development of methane-based rocket engines. In contrast to the well-understood hydrogen-oxygen rocket engines that have been used during the last decades, Lox/Methane imposes additional difficulties on the simulation of the processes within the combustion chamber. For two main reasons the simulation of a methane combustion chamber imposes difficulties at the high-pressure conditions: (1) the relatively complex kinetic mechanisms require the use of tabulated chemistry schemes (2) the close vicinity to the critical conditions for methane and oxygen. Both critical points are located

M. Pini et al. (Eds.): NICFD 2020, ERCOFTAC Series 28, pp. 47–56, 2021.
https://doi.org/10.1007/978-3-030-69306-0_6

around 50 bar and 150 K. Thus, it is unknown whether vapor-liquid-equilibrium effects (VLE) may have to be considered within the thermodynamic description, especially in case for close-critical combustion pressures that are typically present at throttled operating points typical for future engines. These effects were not significant for technical applications of hydrogen-oxygen combustion chambers due to the low critical pressure of hydrogen.

In cryogenic rocket combustion chamber simulations one major challenge is to correctly predict the flow field, the non-linear mixing processes, as well as the combustion of the cryogenic fuels at high-pressure conditions. This requires the estimation of thermal loads onto the combustion chamber walls of future reusable rocket systems by means of validated numerical methods for industrial applications. In the last years effort has been put on the cryogenic mixture modeling for high-pressure hydrogen-oxygen flames. In order to increase the understanding of the physical and thermodynamic effects within methane powered combustion chambers, detailed numerical and experimental investigations are needed. In many cases the simulation of methane-oxygen is still challenging and only few reference test cases are available in the open literature.

One example is the seven injector gaseous combustor of the TU Munich which was considered by many numerical groups around the world (see, e.g. the review paper of Periakis et al. [8]). However, this test case is limited to gaseous fuels and requires the simulation of a full combustion chamber with limited experimental data for comparison. In our present attempt we aim to extend the simple two-dimensional test case introduced by Ruiz et al. [10] as a numerical benchmark for liquid methane-oxygen mixing and combustion processes. In order to keep the geometry simple, a two-dimensional test case focusing on a cut through a coaxial injector is selected. This models the first part of the jet breakup and mixing. It is evident that due to the inherent three-dimensional structure of turbulence this only allows to conduct numerical sensitivity studies and comparisons. Nevertheless, it is a useful test case for code validation as well as case studies due to its simplicity that allows us to use scale resolving methods for detailed comparisons as well.

In the first part of the publication we describe the scope of the research that includes a short summary of the numerical methods used as well as the EOS approximation and the combustion modeling. This is followed by a description of the test case with the first test results. The publication is concluded with a summary and an outlook on further research activities.

2 Scope of Research

In the scope of this research we have a close look at the thermodynamic processes appearing at the cryogenic interface between the cold liquid oxygen jet and the (gaseous) methane jet at supercritical conditions. Similar to the previous investigations in the original test case proposed by Ruiz et al. [10] with hydrogen/oxygen we choose a (constant) combustion chamber pressure of 100 bar, which is located far in the supercritical part to avoid that mixture VLE effects have an influence.

The baseline case will focus on the similarities between the methane-oxygen and hydrogen-oxygen mixing processes. Here combustion kinetics will not be included. For both fuel/oxidizer combinations the fuel is injected as gas while the oxidizer is cryogenic. For the reference case using the combination hydrogen and oxygen, detailed numerical data is available for comparison (see e.g. [7,10]).

In a second step this case will be extended and a chemical kinetics model for high-pressure combustion of methane will be used. One suitable reaction mechanism is the reduced scheme published by Zhukov and Kong [12] that was already used successfully in other rocket chamber applications. This is a first step towards the inclusion of all features of a rocket combustion chamber allowing a detailed comparison of numerical and thermodynamic models. Due to the simple model configuration, parametric studies are feasible.

3 Methodology

For the simulation of cryogenic reacting mixture we are using the DLR flow solver TAU together with suitable EOS mixture models for the cryogenic states. In the following parts we describe shortly the relevant flow solver parts used in this study.

3.1 Compressible Flow Solver TAU

The baseline ideal gas flow solver is the DLR TAU Code for the compressible Navier-Stokes equations with various Reynolds Averaged Navier-Stokes (RANS) and Detached Eddy Simulation (DES) turbulence models. Only a brief overview is given here, detailed descriptions of the ideal-gas flow solver can be found in Hannemann [3], and Karl [5]. TAU is a hybrid grid, finite volume, compressible flow solver. It has been verified for a variety of steady and unsteady flow cases, ranging from sub- to hypersonic Mach numbers.

In recent years the modeling of the flow solver has been extended to cope with cryogenic flows that are e.g. present in rocket combustion chambers (see e.g. in Fechter et al. [1,2]). For these applications the ideal-gas flow solver was extended to cope with cryogenic fluid mixtures based on tabulation as well as direct evaluation of cubic EOS mixture models. The mixture transport properties are chosen in according to the high-pressure Chung model (as described in [9]).

In addition to the cryogenic thermodynamic handling the kinetic modeling was extended to cope with complex schemes based on a cryogenic flamelet model (see Horchler et al. [4]).

3.2 Cryogenic Thermodynamics

To be able to describe the cryogenic mixture effects we apply a cubic EOS mixture model based on a generalized cubic description. Within the cubic EOS formulation the mixture pressure is described by

$$p = \frac{\rho R_u T}{M_w - b\rho} - \frac{a\alpha(T)\rho^2}{(M_w + \delta_1 b\rho)(M_w + \delta_2 b\rho)} \tag{1}$$

using the cubic EOS mixture coefficients $a\alpha(T)$ and b. The EOS parameters δ_1 and δ_2 can be chosen such that the cubic Peng-Robinson or Soave-Redlich-Kwing cubic EOS (see e.g. Kim et al. [6]) are obtained. Corresponding estimation methods for the transport properties are chosen such that they are valid at these high-pressure conditions, see e.g. the mixture viscosity approach of Chung as described in Poling et al. [9].

It is still an open question is whether non-ideal mixture effects close to the critical point region may need to include vapor-liquid equilibrium calculations to be able to numerically reproduce the correct phase informations. First studies on that topic have been done e.g. by Traxinger et al. [11] comparing simulations to simple experiments for a case without chemical kinetics.

3.3 Combustion Modeling Based on the Flamelet Approach

As methane combustion mechanisms typically consist of many reactions, we follow a Flamelet tabulation approach (for general information about the realgas flamelet modeling see Horchler et al. [4]) that is suitable for the application to combustion processes at cryogenic conditions. We solve the modified set of Flamelet equations [6]

$$\begin{aligned} -\rho\frac{\chi}{2}\frac{\partial^2 Y_s}{\partial Z^2} &= \dot{m}_s \\ -\rho\frac{\chi}{2c_p}\left(\frac{\partial^2 h}{\partial Z^2} - \sum_{s=1}^{N_s} h_s \frac{\partial^2 Y_s}{\partial Z^2}\right) &= -\frac{1}{c_p}\sum_s h_s \dot{m}_s \end{aligned} \tag{2}$$

for different values of the scalar dissipation rate χ. The reduced methane reaction mechanism of Zhukov and Kong [12] is used to tabulate the flamelets in mixture fraction space and provided as a flamelet lookup-table to the compressible flow solver.

4 Test Case Definition

This testcase extends the simple test case introduced by Ruiz et al. [10] for cryogenic liquid oxygen and hydrogen to a new application within the methane environment. In the past several studies were conducted on cryogenic mixing effects of hydrogen and oxygen (see e.g. [7,10]) and the original test case was valuable for the validation of numerical tools for cryogenic mixtures in rocket combustion chambers. A sketch of the numerical domain as well as the boundary conditions of the new test case can be found in Fig. 1. For the test case we maintain a oxidizer to fuel ratio of 2.95, a typical value for methane combustion chambers. The methane combustion chamber experiment of the TU Munich (see e.g. Periakis et al. [8]) used a similar oxidizer to fuel ratio.

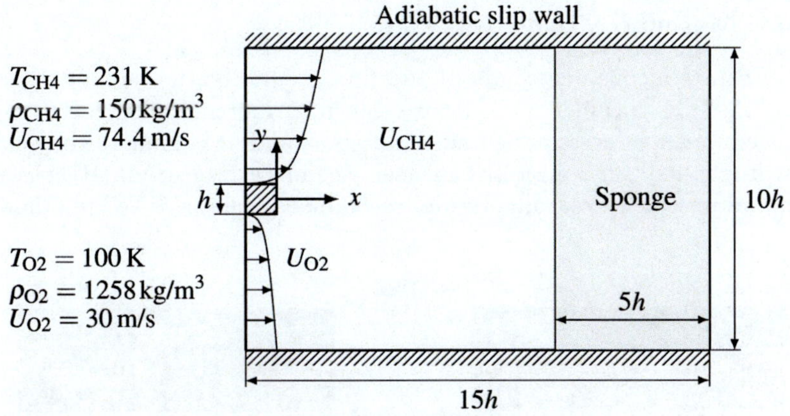

Fig. 1. Sketch of the new mixing test case for Methane/Oxygen following the geometrical definition of the original test case proposed by Ruiz et al. [10].

4.1 New Methane Mixing Test Case

With the increased interest in the modeling of methane, we decided to set up a similar reference test case at conditions typical for methane/oxygen flames at cryogenic conditions. For methane, supercritical conditions are chosen while the oxidizer is injected at a subcritical temperature. At the inlets we impose a $1/7$ power law velocity profile to account for the upstream tubing prior to the injector.

4.2 New Methane Reacting Test Case

As an extension of the mixing test case, we consider the same geometry as in the previous case with an additional combustion model. Due to the large combustion kinetics of methane flames, we apply a realgas flamelet approach. The used kinetics scheme is based on the methane mechanism published by Zhukov and Kong [12] that is suitable for high-pressure methane combustion.

5 First Results

In this paper we include the first results for the Ruiz mixing and reacting test case. A short discussion about the similarities between the original Ruiz test case with fluid pairing hydrogen-oxygen with the proposed new one using methane-oxygen is added.

The authors would like to mention that the direct comparison of RANS and two-dimensional DES/LES solutions is rather quantitative and illustrative. One would expect that a comparison of the two-dimensional RANS solution to a three-dimensional LES simulation would yield different results.

5.1 New Methane Mixing Test Case

In Fig. 2 instantaneous snapshots of the first results of the new mixture test case using the SST turbulence model are shown. A more detailed comparison of the averaged mean values can be found in Figs. 3 and 4. Similar to the outcome of the hydrogen-oxygen Ruiz test case (see e.g. in Fechter et al. [1]), the mean turbulent values can be reasonably well reproduced using a RANS method.

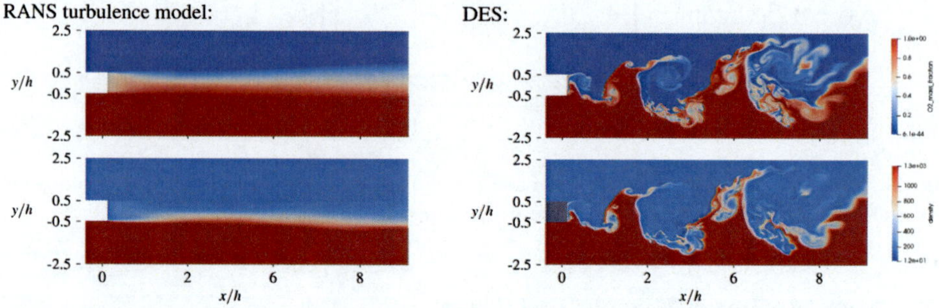

Fig. 2. First results of the new two-dimensional LOX/CH4 mixing case: Left: Averaged fields of oxygen mass fraction and density using the RANS SST turbulence model; Top: Instantaneous snapshots of the oxygen mass fraction and density using a DES turbulence model.

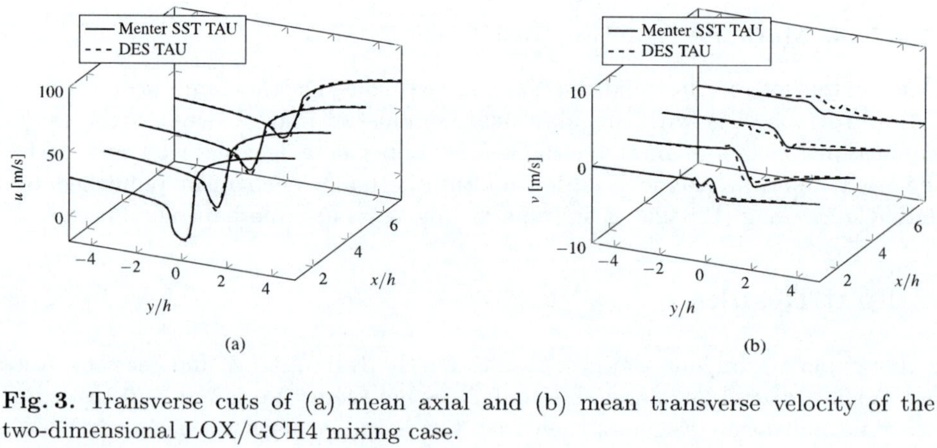

Fig. 3. Transverse cuts of (a) mean axial and (b) mean transverse velocity of the two-dimensional LOX/GCH4 mixing case.

The major differences to the original Ruiz test case [10] using LOX/GH2 is the higher density of the methane at the instream conditions; the oxygen

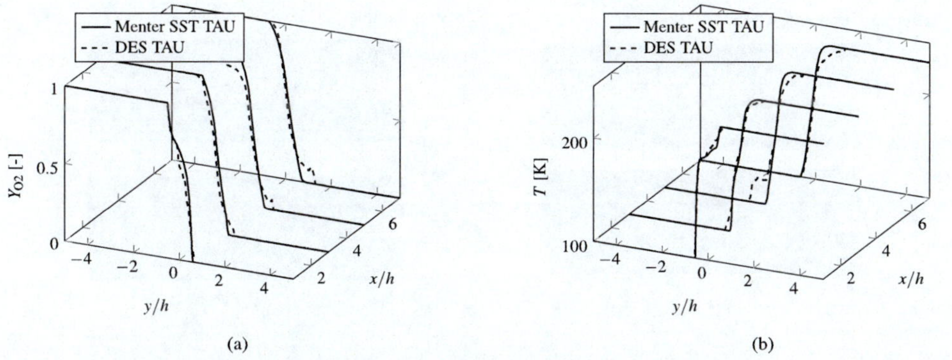

Fig. 4. Transverse cuts of (a) mean oxygen mass fraction and (b) mean temperature of the two-dimensional LOX/GCH4 mixing case.

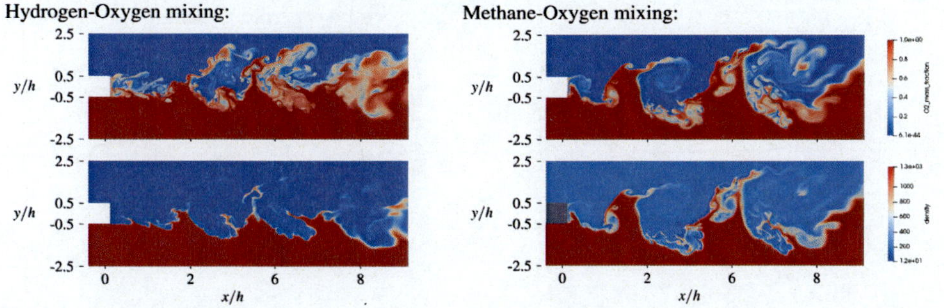

Fig. 5. Comparison of the vortex structures developed in the original Ruiz test case with the fluid pairing hydrogen-oxygen (left) with the new test case for methane-oxygen mixing (right).

conditions are kept unchanged (see comparison of instantaneous snapshots in Fig. 5). Due to the different material properties of the fuel, the Reynolds number for the mixing layer is about twice as large as for the methane case,

$$\mathrm{Re}_{\mathrm{O_2H_2}} = \frac{\rho v h}{\nu} = 164 \cdot 10^6 \quad ,$$
$$\mathrm{Re}_{\mathrm{O_2CH_4}} = 344 \cdot 10^6 \quad .$$

This increased Reynolds number in the methane test case implies that the small-scale turbulent structuresm that can be identified in the hydrogen-oxygen test case, can not be resolved on the used DES grid. A finer grid would be needed for the methane test case to resolve the same amount of small-scale turbulence structures for the methane mixing test case.

5.2 New Methane Reacting Test Case

In Fig. 6 instantaneous snapshots of the first results of the new reacting test case are shown using the RANS SST turbulence model as well as a DES turbulence

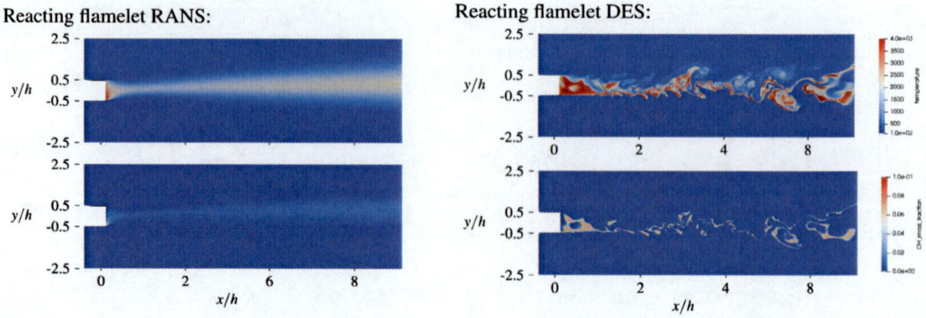

Fig. 6. Comparison of the vortex structures developed in the hydrogen-oxygen reacting test case (left) with the new methane-oxygen reacting test case (right).

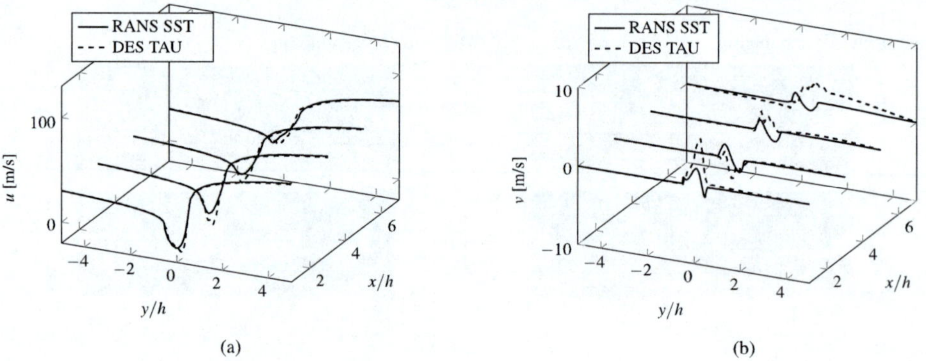

Fig. 7. Transverse cuts of (a) mean axial and (b) mean transverse velocity of the two-dimensional LOX/GCH4 reacting case.

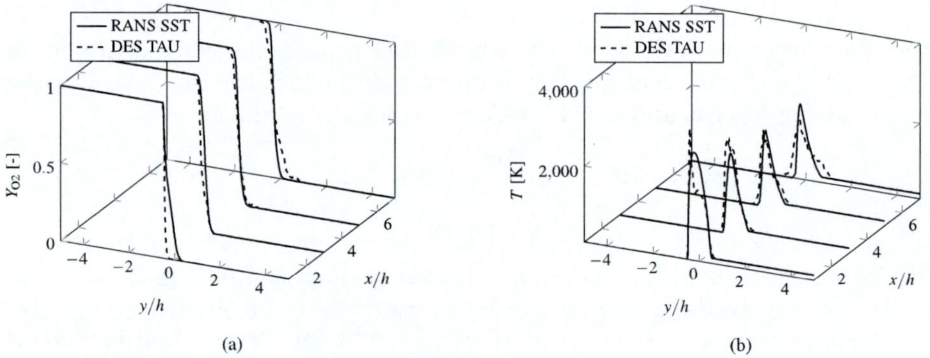

Fig. 8. Transverse cuts of (a) mean oxygen mass fraction and (b) mean temperature of the two-dimensional LOX/GCH4 reacting case.

model. To model the combustion we apply a Flamelet combustion model with the reduced reaction kinetics described in Zhukov and Kong [12]. The combustion effects are limited to a thin layer inbetween the methane and oxygen streams.

A detailed comparison of the flow features using the averaged mean values can be found in Figs. 7 and 8. Similar to the mixing case, the mean turbulent values can be reasonably well reproduced using the RANS approach. In comparison to the mixing test in Fig. 2 the vortex structures for the reacting case in Fig. 6 are much smaller in size. This is due to the influence of the reactions and the additional temperature gradient in the shear layer.

6 Conclusions and Future Work

We define a generic two-dimensional test case for the mixture and combustion processes of cryogenic methane and oxygen. This simplified two-dimensional numerical test case is inspired by a cut through a coaxial injector and therefore suitable for the validation of numerical methods that aim to describe the physical and thermodynamic effects within the simplified injector geometry. Due to the simplified geometry it is suitable to apply higher scale resolving turbulence modeling approaches like DES or LES in order to tune RANS based turbulence models for this type of applications. This allows us to produce valuable validation data for turbulence models at these cryogenic conditions.

Here we presented the first results for the mixing and reacting test case, comparing the outcome of RANS models to higher resolved DES methods. A first comparison shows a good agreement between the two methods for the mean values. In comparison to the original test case based on hydrogen-oxygen mixing, the methane-oxygen test case shows more turbulent structures in the mixing layer due to the higher Reynolds number and different fluid properties. To resolve these turbulent structures a finer DES mesh resolution is needed. In future this test case might serve as example to investigate the effect of flame anchoring in methane flames.

Acknowledgments. The present work was conducted in the framework of the German Aerospace Center (DLR) project TAUROS (TAU for Rocket Thrust Chamber Simulation) and AMADEUS (Advanced Methods for Reusable Aerospace Vehicle DEsign Using Artificial Intelligence and Interdisciplinary Numerical Simulation) focusing on the qualification and advancement of the DLR flow solver TAU for liquid rocket thrust chamber applications. The financial support of the DLR Space Research Programmatic is highly appreciated.

References

1. Fechter, S., Horchler, T., Karl, S., Hannemann, K.: Investigation of rans turbulence models for cryogenic rocket combustion chambers. In: 8th European Conference for Aeronautics and Space Sciences (EUCASS) 2019, Madrid, Spain (2019)
2. Fechter, S., Horchler, T., Karl, S., Hannemann, K., Suslov, D., Hardi, J., Oschwald, M.: Efficient handling of cryogenic equation of state for the simulation of rocket combustion chambers. In: 2nd International Seminar on Non-Ideal Compressible-Fluid Dynamics for Propulsion & Power, Bochum, Germany (2018)

3. Hannemann, V.: Numerische Simulation von Stoß-Stoß-Wechselwirkungen unter Berücksichtigung von chemischen und thermischen Nichtgleichgewichtseffekten. Ph.D. thesis (1997)
4. Horchler, T., Fechter, S., Karl, S., Hannemann, K.: Comparison of detailed chemistry and flamelet combustion modeling in a H2/LOx subscale combustion chamber. In: 17th International Conference on Numerical Combustion 2018, May 2019
5. Karl, S.: Numerical investigation of a generic scramjet configuration. Ph.D. thesis, University of Dresden (2011)
6. Kim, S.-K., Choi, H.-S., Kim, Y.: Thermodynamic modeling based on a generalized cubic equation of state for kerosene/LOx rocket combustion. Combust. Flame **159**(3), 1351–1365 (2012)
7. Ma, P.C., Lv, Y., Ihme, M.: An entropy-stable hybrid scheme for simulations of transcritical real-fluid flows. J. Comput. Phys. **340**, 330–357 (2017)
8. Perakis, N., Haidn, O.J., Eiringhaus, D., Rahn, D., Zhang, S., Daimon, Y., Karl, S., Horchler, T.: Qualitative and quantitative comparison of rans simulation results for a 7-element GOX/GCH4 rocket combustor. In: 2018 Joint Propulsion Conference, p. 4556 (2018)
9. Poling, B.E., Prausnitz, J.M., O'Connel, J.P.: The Properties of Gases and Liquids, 5th edn. McGraw Hill Book Co., New York (2007)
10. ARuiz, A.M., Lacaze, G., Oefelein, J.C., Mari, R., Cuenot, B., Selle, L., Poinsot, T.: Numerical benchmark for high-Reynolds-number supercritical flows with large density gradients. AIAA J. **54**, 1445–1460 (2015)
11. Traxinger, C., Müller, H., Pfitzner, M., Baab, S., Lamanna, G., Weigand, B., Matheis, J., Stemmer, C., Adams, N.A., Hickel, S.: Experimental and numerical investigation of phase separation due to multi-component mixing at high-pressure conditions. arXiv preprint arXiv:1706.03923 (2017)
12. Zhukov, V.P., Kong, A.F.: A compact reaction mechanism of methane oxidation at high pressures. Prog. React. Kinet. Mech. **43**(1), 62–78 (2018)

Numerical Characterization of Premixed Methane Flames in Vitiated Atmosphere at Supercritical Conditions

Federico Lo Presti[1(✉)], Pascal Post[1], Francesca di Mare[1], and Jeroen van Oijen[2]

[1] Ruhr University Bochum, Bochum, Germany
{federico.lopresti,pascal.post,francesca.dimare}@rub.de
[2] Eindhoven University of Technology, Einhoven, The Netherlands
J.A.v.Oijen@tue.nl

Abstract. Some cutting edge technologies for power generation make use of Supercritical CO_2 cycles to improve efficiency and reduce emissions. While all of these technologies are characterized by non-premixed combustion conditions, premixed flames at very high pressure are an almost unexplored field, being concerns on their stability a strong limiting factor. In this study, premixed methane flames in a carbon dioxide atmosphere have been characterized numerically with a one dimensional detailed chemistry solver. Their distinctive characteristics such as flame speed and thickness have been determined at pressures from 100 to 300 bar and at varying CO_2 concentration levels, as a first, fundamental step towards the characterization of their stability limits.

Keywords: Premixed flames · Supercritical CO_2 · Supercritical combustion · High pressure · Detailed chemistry

1 Introduction

Combustion in devices such as liquid rocket engines and diesel engines is often designed to happen at very high pressure to allow the most efficient operation. Temperature and pressure of the fluids involved can have values which are across or higher than their critical ones. In this conditions, fluids are in the so called transcritical or supercritical state, where real gas effects on both mixing and combustion processes are of significant importance. Real gas effects consist in abrupt changes of thermodynamic and transport properties such as mixture density, enthalpy, heat capacity, viscosity and a vanishing surface tension, which eliminates the liquid-gas sharp interface [35]. Although this might seem to be a simplification since it removes the need for a two-phase flow treatment, a founding pillar of the standard numerical treatment of fluid flows is taken down: the equations of state (EOS) for the subcritical state lose their validity at this conditions and need to be modified with the inclusion of real gas effects. Also, real

M. Pini et al. (Eds.): NICFD 2020, ERCOFTAC Series 28, pp. 57–68, 2021.
https://doi.org/10.1007/978-3-030-69306-0_7

gas thermodynamic functions such as enthalpy, heat capacity and acoustic speed, need to be derived from the new equation of state and transport properties, which are usually modelled with simple but effective laws (e.g. Sutherland), need a more accurate modelling that takes into account the effects of inter-molecular forces [19]. The importance of real gas effects has been underlined in numerous works [4,5,19,26,32]. For many applications, supercritical combustion happens in the non-premixed regime. There is a wide range of studies of counterflow diffusion flames configurations that investigate the laminar non-premixed flame structure at supercritical conditions. In [19] hydrogen–oxygen diffusion flames are analysed to determine the effect of parameters such as pressure, strain rate and inlet temperature on the flame. Extinction limits of the same kind of supercritical diffusion flames as well as the influence of pressure on the heat release have been studied in [18]. Hydrocarbons mixtures have been investigated in [36] as well as the effect of fuel composition in [20]. To improve computational efficiency, the flamelet method [29,30] is preferred to the direct solution of a conservation law for each chemical species involved in the combustion process when a time and length scale separation of turbulence and chemistry can be assumed. With regards to supercritical combustion, the method has been proven to be valid in [40] and employed in successive works [19,22]. In [9] the flamelet method is employed in the development of a model for the simulation of reacting mixtures of real gases. In [39] a tabulated method based on the flamelet assumption is developed for LES applications. Alongside the applications mentioned above and the big trend in the combustion of hydrogen, interest is growing in the development of supercritical carbon dioxide (sCO_2) combustors, motivated by the increased heat-to-electricity conversion efficiencies and higher power density [24] if compared with steam-based power cycles. Dilution of fuel-air mixture by recirculation of burned gases is also more commonly employed to reduce NOx emissions, thanks to lower peak combustion temperatures, and to improve flame stability [21]. A deeper understanding of the dynamics and structure of CO_2-diluted premixed flames at supercritical conditions is therefore of significant importance for the further development of such technologies. Numerical investigations have been performed on combustors [6] and on jets [8] of methane in a supercritical CO_2 environment. In the context of premixed flames at supercritical conditions, an important aspect to be investigated are the stability mechanisms in relation to geometrical and boundary conditions, starting from the considerations of premixed flames at low pressures done in [10] and [38]. Adebiyi et al. [2] investigated the propagation characteristics of premixed laminar flames in Oxy-Fuel combustion at supercritical conditions in a CO_2-atmosphere, however they used a single-step Arrhenius-type chemistry. A further application where CO_2-diluted flames can be found is the combustion of biogas obtained from anaerobic digestion of organic matter. In this context, there have been many works [3,7,11,14–17,27], both experimental and numerical, investigating the laminar flame properties such as flame speed and thickness in CO_2-diluted flames with varying fuel dilution grade, equivalence ratios, unburnt mixture temperature and pressure conditions. In a recent review of these works [37], the fundamental findings for laminar flames of methane/air/carbon dioxide and propane/air/carbon dioxide mixtures are summarized. The pressure conditions examined to date, however, do not exceed the 10 bar and it might be interesting to investigate the effects of very high pressure on this kind of flames.

2 Scope of Research

Within this work, numerical calculations of premixed methane flames with various amounts of CO_2 in the premixed mixture, at very high pressures, have been performed. Starting point will be the calculation of biogas-like mixtures, for which flame speed, thickness and temperatures will be examined and the trend by varying CO_2-dilution grade will be compared with results obtained at atmospheric condition. Next, premixed flames of mixtures typical of oxy-fuel combustion with varying CO_2 concentration will be analysed. With regards to the numerical methods, the influence of the equation of state and of the reaction mechanism will be investigated. These results should lay the foundations for a subsequent application to turbulent flows as done in [9,22], where, however, essentially diffusion flames have been taken into account. An FGM approach [34] will be developed, for which one-dimensional calculations in the physical space at various levels of equivalence ratio and/or different temperature of the unburnt mixture are performed and parametrized as a function of some progress variable and mixture fraction to generate a lookup table to be used in combination with CFD solvers.

3 Methodology

The work is carried out with the one dimensional solver CHEM1D [1] where the underlying ideal gas model has been replaced by a real-gas ansatz allowing a description of the combustion manifolds consistent with the thermodynamic model adopted for flow solution. The solver has been extended with the Peng-Robinson [28] equation of state and the calculations of enthalpy and isobaric heat capacity have been modified consistently with the new equation of state. The calculation of transport coefficients (viscosity and thermal conductivity) has been adapted to high pressure conditions with the introduction of Chung's rules for mixtures at high pressure which are, despite the scarceness of experimental data for the calculation of the correlation coefficients, one of the best analytical approximations available [31]. The importance of the real gas models (equation of state, thermodynamics and transport), has been systematically analysed in [13] for hydrodynamic flames where supercritical water is present. The conclusions drawn in this work state that the most important modification to the model is the equation of state, which has a major influence on the flame propagation speed. A key aspect when dealing with detailed chemistry computations is the choice of the reaction mechanism. Within this work, the widely used GRI 3.0 reaction mechanism [33] will be compared with the Aramco 2.0 [25] mechanism as well as with a reduced version of the latter. The first mechanism, comprising

53 species and 255 reactions, has been extensively validated for the combustion of hydrocarbons for pressures up to 10 atm but its performance in the high pressure range is questionable. The Aramco2.0 mechanism comprises 493 species and 2716 reactions and has been validated for methane combustion up to 260 atm. In [23], the two mechanisms are compared for the prediction of Ignition Delay Time in $CH_4/O_2/CO_2$ mixtures at pressures up to 300 atm and Aramco2.0 mechanism seems to yield superior results. In the present context, however, a major drawback of the Aramco2.0 mechanism is the huge number of species which are taken into account. In fact, the calculation of mixture averaged critical properties, needed for both the iterative calculation of density with PR-EOS as well as for the calculation of transport coefficients, involves repeated double loops on the species so that the computational time is proportional to the square of the number of species and the time needed for the calculation of a single flamelet increases by order of magnitudes when combining the real gas model with this mechanism. To overcome this limitation, a reduced mechanism targeted to the present application has been calculated with a computational tool [12] using the method of Global Pathway Selection to select the most relevant species and reactions. The resulting mechanism contains 37 species and 223 reactions, and will be also compared to GRI3.0 and Aramco2.0 mechanisms.

4 Results and Discussion

4.1 Validation at Atmospheric Conditions

CO_2-diluted flames have been first analysed at atmospheric conditions in order to perform comparisons with available experimental data. The composition of the mixtures is listed in Table 1. The analysis has been restricted to stoichiometric mixtures. Pressure is set to 1 bar and the temperature of the unburnt mixture is $T_u = 300\,\mathrm{K}$.

Table 1. Unburnt mixture composition (molar fractions)

	Fuel		Oxidizer		
Φ	CH_4	CO_2	N_2	O_2	Ar
1.0	1.0	0.0	0.781	0.21	0.009
1.0	0.8	0.2	0.781	0.21	0.009
1.0	0.6	0.4	0.781	0.21	0.009
1.0	0.4	0.6	0.781	0.21	0.009

Figure 1a shows the temperature profile along the one dimensional flame, which is almost identical at this conditions for the three reaction mechanisms. In Fig. 1b, the laminar flame speed has been compared with experimental data: a good agreement is achieved and the values calculated in the present work

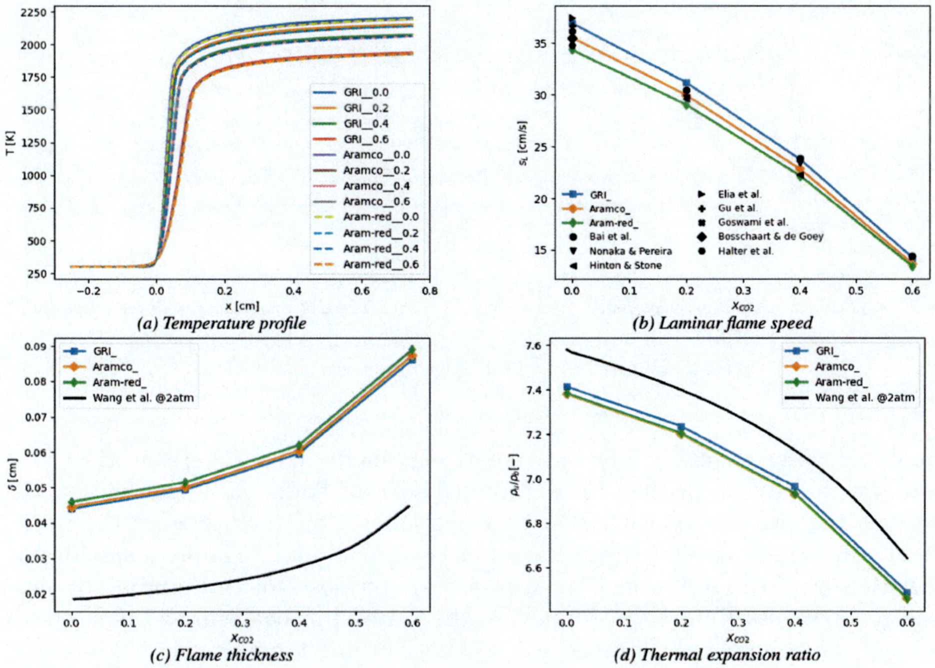

(a) Temperature profile *(b) Laminar flame speed*

(c) Flame thickness *(d) Thermal expansion ratio*

Fig. 1. Flamelets at 1 bar, comparison with experiments [3,7,11,15–17,27] extracted from [37]

lie in the range of experimental results. Figures 1c and 1d show respectively the variation with CO_2 dilution of flame thickness and thermal expansion ratio, defined as the ratio between unburnt mixture and burnt mixture densities. Lines do not coincide because of the slightly different pressure in the results from [37], although the trend is clearly the same. The findings reported in the literature are here all reproduced, in particular the lower temperature of the flame, lower speed and higher thickness with the increase in dilution of fuel with CO_2 .

4.2 High Pressure

The analysis has been repeated for the mixtures defined in Table 1 at the pressures of 100 and 200 bar, with the different reaction mechanisms and with both the standard ideal gas model and the real gas formulation.

Figure 2a and 2b show the comparison of the temperature profiles in the physical space at 200 bar respectively for different reaction mechanisms (Fig. 2a) and different EOS and transport modelling (Fig. 2b). In Fig. 2a, although the temperatures of reaction products are the same, differences in temperature profiles can be noticed between the GRI and Aramco mechanisms, as a results of the higher pressure. For what has been said in the introduction about these mechanisms, it could be argued that results obtained with the Aramco mechanisms

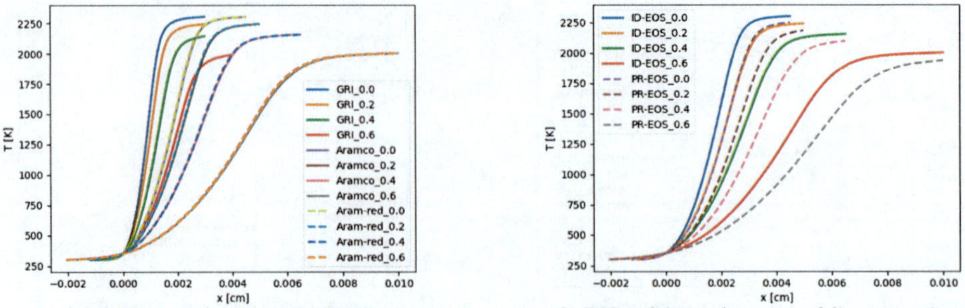

(a) Reaction mechanism comparison *(b) EOS and thermodynamic modeling comparison*

Fig. 2. Flamelets at 200 bar, mixture of Table 1

should be closer to reality. For as much as regards its reduced version, the solutions can hardly be distinguished. Comparisons of flame thickness and speed between the two mechanisms, which are not shown for brevity, gave the same result. This confirms that the reduction carried over doesn't imply a significant deviation from the full model, at least for the purposes of this study. For this reason, as discussed in the introduction, the reduced mechanism will be used in place of the full one and all plots that will be shown in the following sections are obtained with the reduced Aramco2.0 mechanism. The comparison of the gas models (Fig. 2b), obtained using the reduced Aramco2.0 mechanism, confirms the expected deviations from the ideal behaviour, especially in the development of the temperature field.

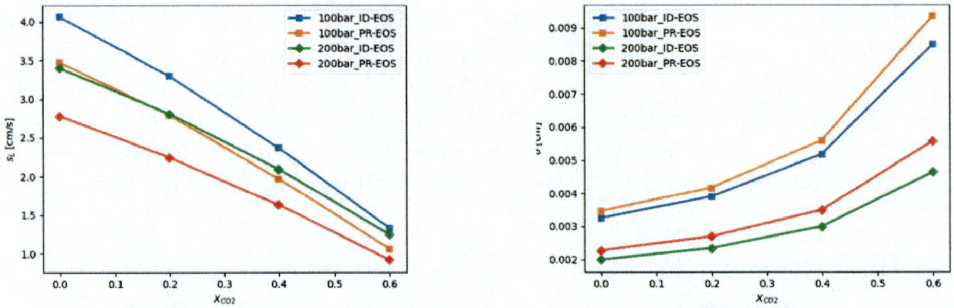

(a) Flame speed *(b) Flame thickness*

Fig. 3. Flamelets at 100 and 200 bar, mixture of Table 1

In Fig. 3, a comparison of flame speed and thickness between the two gas models is shown. For both models, the same trends of the atmospheric results are reproduced. As seen in Fig. 2b, the real gas modelling results in a higher thickness and lower flame speed and differences at this pressure level are not negligible.

4.3 Supercritical

Oxy-Fuel flames with various level of CO_2 at different pressures have been analysed. The compositions of the mixtures are listed in Table 2. The temperature of the unburnt mixture is set to $T_u = 300\,\text{K}$ and flamelets at a pressure level of 100, 200 and 300 bar have been calculated. With this mixture composition and at this conditions, some states of the flame exceed critical values for both pressure and temperature.

In Fig. 4, the states of the mixture resulting from the calculation of the flamelet solutions as a function of reduced pressure and reduced temperature are shown. The grey lines are isolines of compressibility factor Z, where values

Table 2. Unburnt mixture composition (molar fractions)

	Fuel	Oxidizer	
Φ	CH_4	CO_2	O_2
1.0	1.0	0.2	0.8
1.0	1.0	0.4	0.6
1.0	1.0	0.6	0.4
1.0	1.0	0.8	0.2

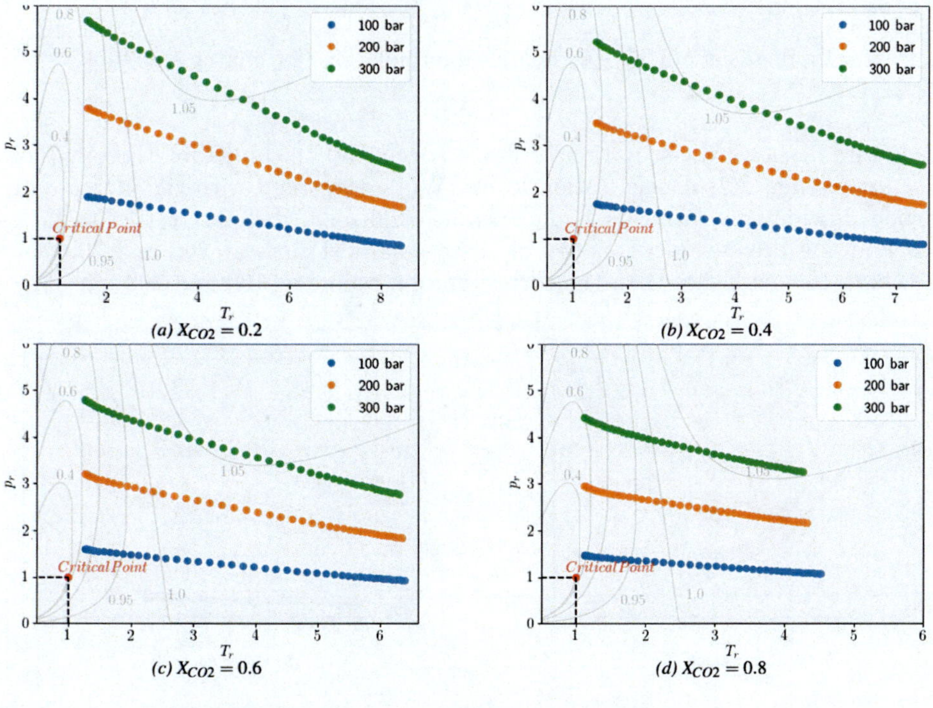

Fig. 4. Flamelets solution in T_r-p_r diagram, mixtures of Table 2

of Z different from one are an indicator of the degree of departure from ideality of the mixture. We see that the points corresponding to lower temperature values (unburnt mixture) have compressibility factor values much smaller than one with pressure and temperature higher than their critical ones. Figure 5 shows the relevant change in calculated density and thermal conductivity with the Peng-Robinson EOS and Chung's rules for transport properties for the mixture with the highest content of CO_2 ($X_{CO2} = 0.8$) at the different values of pressure. Values of density are up to two times larger than the corresponding ideal value, while differences in the values of thermal conductivity are even higher and we observe slopes changes in the transition from the supercritical to the ideal behaviour.

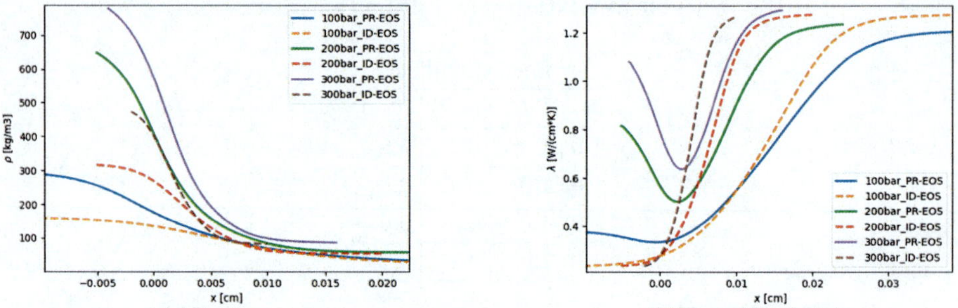

(a) Density *(b) Thermal Conductivity*

Fig. 5. Comparison of ideal and non ideal formulations for mixtures of Table 2

Figure 6 shows the variation of flame speed and thickness for the oxy-fuel case at varying CO_2 concentration levels. While the trend with the increase of carbon dioxide concentration is once again confirmed, it is interesting to look closer at the pressure dependency of these results. Points at 200 and 300 bar are closer to each other than the corresponding points at 100 and 200 bar. The

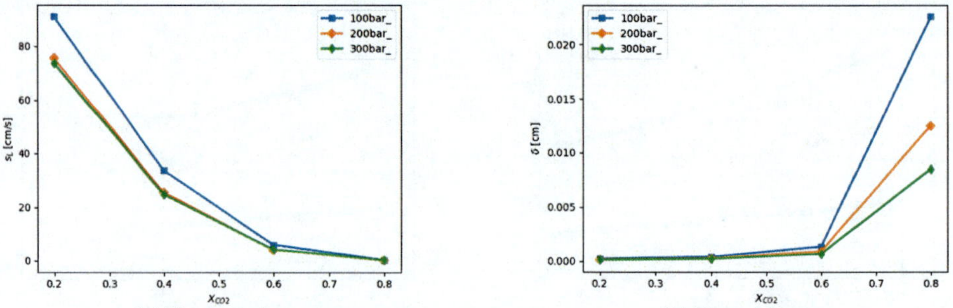

(a) Flame speed *(b) Flame thickness*

Fig. 6. Flamelets at 100, 200 and 300 bar, mixtures of Table 2 pressure dependency of flame thickness is better seen in Fig. 7, where the flame thickness is shown normalized by the value at 100 bar and as a function of pressure. We observe that the trend is between $1/p$ and $1/\sqrt{p}$ and changes with CO_2 concentration. Moreover, the curve for $X_{CO2} = 0.8$ seems to be closer to $1/p$, which is anomalous if compared with the other curves.

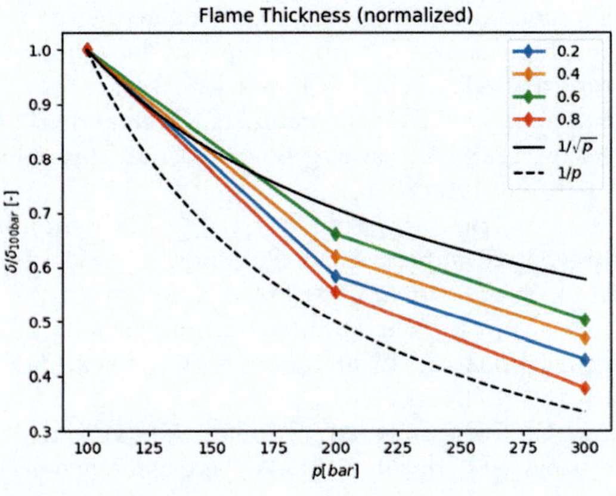

Fig. 7. Normalized flame thickness as a function of pressure

5 Conclusions and Future Works

The calculations performed in this work are a first numerical characterization with detailed chemistry of premixed flames for the combustion of supercritical mixtures with high levels of CO_2 dilution. The lack of experimental data for comparison at higher pressures doesn't allow to draw conclusions on the accuracy of calculations, although the high efforts in the accurate modelling suggest that a certain degree of closeness to reality could be reached. The comparison between the ideal and the real gas approaches have been shown to underline the importance of taking into account real gas effects. Reaction mechanisms also proved to play a very important role. The calculations will be extended to draw continuous lines at varying pressures and carbon dioxide dilution levels, so that tendencies which are not clear with a limited number of points can be better recognized. The one dimensional calculations here performed, also, will be the base for the construction of manifolds to be used in two and three dimensional calculations in combination with a CFD solver.

References

1. CHEM1D: A one-dimensional laminar flame code, Eindhoven University of Technology

2. Adebiyu, A.A., Kemenov, K., Akkerman, V.: Numerical simulation of oxy-fuel premixed combustion in a supercritical CO_2-diluted environment. Proc. West Virginia Acad. Sci. **90**(1) (2018)
3. Bai, Z., Wang, Z., Yu, G., Yang, Y., Metghalchi, H.: Experimental study of laminar burning speed for premixed biomass/air flame. J. Energy Resour. Technol. **141**(2), 022206 (2019)
4. Banholzer, M., Pfitzner, M., Sakellarakis, D., Wright, Y.M., Vera-Tudela, W., Boulouchos, K.: Numerical investigation of nozzle-geometry variations and back-pressure changes on high pressure gas injections under application-relevant conditions. Technical report, SAE Technical Paper (2018)
5. Banuti, D.T., Hannemann, V., Hannemann, K., Weigand, B.: An efficient multi-fluid-mixing model for real gas reacting flows in liquid propellant rocket engines. Combust. Flame **168**, 98–112 (2016)
6. Banuti, D.T., Shunn, L., Bose, S., Kim, D.: Large Eddy Simulations of Oxy-Fuel combustors for direct-fired supercritical CO_2 power cycles. In: The 6th International Symposium on Supercritical CO_2 Power Cycles (2018)
7. Bosschaart, K.J., De Goey, L.: The laminar burning velocity of flames propagating in mixtures of hydrocarbons and air measured with the heat flux method. Combust. Flame **136**(3), 261–269 (2004)
8. Chong, S.T., Tang, Y., Hassanaly, M., Raman, V.: Turbulent mixing and combustion of supercritical jets. In: 55th AIAA Aerospace Sciences Meeting, p. 0141 (2017)
9. Cutrone, L., De Palma, P., Pascazio, G., Napolitano, M.: A flamelet/progress-variable approach for the simulation of turbulent combustion of real gas mixtures. In: 44th AIAA/ASME/SAE/ASEE Joint Propulsion Conference & Exhibit, p. 4567 (2008)
10. De Goey, L., Van Oijen, J., Kornilov, V., ten Thije Boonkkamp, J.: Propagation, dynamics and control of laminar premixed flames. Proc. Combust. Inst. **33**(1), 863–886 (2011)
11. Elia, M., Ulinski, M., Metghalchi, M.: Laminar burning velocity of methane-air-diluent mixtures. J. Eng. Gas Turbines Power **123**(1), 190–196 (2001)
12. Gao, X., Yang, S., Sun, W.: A global pathway selection algorithm for the reduction of detailed chemical kinetic mechanisms. Combust. Flame **167**, 238–247 (2016)
13. Gao, Z., Wang, H., Song, C., Luo, K., Fan, J.: Real-fluid effects on laminar diffusion and premixed hydrothermal flames. J. Supercrit. Fluids **153**, 104566 (2019)
14. Goswami, M., Derks, S.C., Coumans, K., Slikker, W.J., de Andrade Oliveira, M.H., Bastiaans, R.J., Luijten, C.C., de Goey, L.P.H., Konnov, A.A.: The effect of elevated pressures on the laminar burning velocity of methane + air mixtures. Combust. Flame **160**(9), 1627–1635 (2013)
15. Gu, X.J., Haq, M.Z., Lawes, M., Woolley, R.: Laminar burning velocity and Markstein lengths of methane-air mixtures. Combust. Flame **121**(1–2), 41–58 (2000)
16. Halter, F., Chauveau, C., Djebaïli-Chaumeix, N., Gökalp, I.: Characterization of the effects of pressure and hydrogen concentration on laminar burning velocities of methane-hydrogen-air mixtures. Proc. Combust. Inst. **30**(1), 201–208 (2005)
17. Hinton, N., Stone, R.: Laminar burning velocity measurements of methane and carbon dioxide mixtures (biogas) over wide ranging temperatures and pressures. Fuel **116**, 743–750 (2014)
18. Huo, H., Wang, X., Yang, V.: A general study of counterflow diffusion flames at subcritical and supercritical conditions: oxygen/hydrogen mixtures. Combust. Flame **161**(12), 3040–3050 (2014)

19. Lacaze, G., Oefelein, J.C.: A non-premixed combustion model based on flame structure analysis at supercritical pressures. Combust. Flame **159**(6), 2087–2103 (2012)
20. Lapenna, P.E., Ciottoli, P.P., Creta, F.: The effect of fuel composition on the non-premixed flame structure of LNG/LOx mixtures at supercritical pressure. In: 54th AIAA Aerospace Sciences Meeting, p. 0690 (2016)
21. Lefebvre, A.H.: Gas Turbine Combustion. CRC Press, Boca Raton (1998)
22. Ma, P.C., Banuti, D., Hickey, J.-P., Ihme, M.: Numerical framework for transcritical real-fluid reacting flow simulations using the flamelet progress variable approach. In: 55th AIAA Aerospace Sciences Meeting, p. 0143 (2017)
23. Manikantachari, K., Vesely, L., Martin, S., Bobren-Diaz, J.O., Vasu, S.: Reduced chemical kinetic mechanisms for oxy/methane supercritical CO_2 combustor simulations. J. Energy Resour. Technol. **140**(9), 092202 (2018)
24. McClung, A., Brun, K., Chordia, L.: Technical and economic evaluation of supercritical oxy-combustion for power generation. In: The 4th International Symposium-Supercritical CO2 Power Cycles (2014)
25. Metcalfe, W.K., Burke, S.M., Ahmed, S.S., Curran, H.J.: A hierarchical and comparative kinetic modeling study of c1–c2 hydrocarbon and oxygenated fuels. Int. J. Chem. Kinet. **45**(10), 638–675 (2013)
26. Müller, H., Niedermeier, C.A., Matheis, J., Pfitzner, M., Hickel, S.: Large-eddy simulation of nitrogen injection at trans-and supercritical conditions. Phys. Fluids **28**(1), 015102 (2016)
27. Nonaka, H., Pereira, F.: Experimental and numerical study of CO_2 content effects on the laminar burning velocity of biogas. Fuel **182**, 382–390 (2016)
28. Peng, D.-Y., Robinson, D.B.: A new two-constant equation of state. Ind. Eng. Chem. Fundam. **15**(1), 59–64 (1976)
29. Peters, N.: Turbulent Combustion. IOP Publishing, Bristol (2001)
30. Pitsch, H.: Large-eddy simulation of turbulent combustion. Annu. Rev. Fluid Mech. **38**, 453–482 (2006)
31. Reid, R.C., Prausnitz, J.M., Poling, B.E.: The Properties of Gases and Liquids (1987)
32. Sakellarakis, V.D., Banholzer, M., Llugaliu, D., Wright, Y., Pfitzner, M., Boulouchos, K.: Assessment of real-gas effects in high-pressure gas injection at engine-relevant conditions. In: 12th International ERCOFTAC Symposium on Engineering Turbulence Modelling and Measurements (2018)
33. Smith, G.P.: Gri-mech 3.0 (1999). http://www.me.berkley.edu/gri_mech/
34. Van Oijen, J., De Goey, L.: Modelling of premixed laminar flames using flamelet-generated manifolds. Combust. Sci. Technol. **161**(1), 113–137 (2000)
35. Wang, X., Huo, H., Yang, V.: Supercritical combustion of general fluids in laminar counterflows. In: 51st AIAA Aerospace Sciences Meeting including the New Horizons Forum and Aerospace Exposition, p. 1165 (2013)
36. Wang, X., Huo, H., Yang, V.: Counterflow diffusion flames of oxygen and n-alkane hydrocarbons (ch4-c16h34) at subcritical and supercritical conditions. Combust. Sci. Technol. **187**(1–2), 60–82 (2015)
37. Wang, Z., Yelishala, S.C., Yu, G., Metghalchi, H., Levendis, Y.A.: Effects of carbon dioxide on laminar burning speed and flame instability of methane/air and propane/air mixtures: a literature review. Energy Fuels **33**(10), 9403–9418 (2019)
38. Yu, J., Yu, R., Bai, X., Bastiaans, R., Van Oijen, J., De Goey, L.: Heat transfer and flame stabilization of laminar premixed flames anchored to a heat-flux burner. Int. J. Hydrogen Energy **41**(3), 2037–2051 (2016)

39. Zips, J., Müller, H., Pfitzner, M.: Efficient thermo-chemistry tabulation for non-premixed combustion at high-pressure conditions. Flow Turbul. Combust. **101**(3), 821–850 (2018)
40. Zong, N., Ribert, G., Yang, V.: A flamelet approach for modeling of liquid oxygen (LOX)/methane at supercritical pressures. In: Forty-Sixth AIAA Aerospace Sciences Meeting and Exhibit, vol. 946, pp. 1–16. AIAA (2008)

Numerical Methods

Towards Subgrid-Scale Turbulence Modeling in Dense Gas Flows

Alexis Giauque(✉), Aurélien Vadrot, Paolo Errante, and Christophe Corre

LMFA - Laboratoire de Mécanique des Fluides et d'Acoustique Ecole Centrale de Lyon, 36 avenue Guy de Collongue, 69134 Ecully Cedex, France
alexis.giauque@ec-lyon.fr

Abstract. It is known from the literature that Large Eddy Simulations of dense gas flows can prove very useful in order to better understand the behavior of friction losses in Organic Rankine Cycle turbines and expanders. Yet, no dedicated turbulence closure models exist for such flows. In this study, the authors lay the fondations for the development of turbulence closure models in dense gas flows. The Subgrid-Scale terms are first rigorously derived from the filtered Navier-Stokes equations. Additional terms to the well-known Subgrid-Scale closure terms are identified in the dense gas context. Using Direct Numerical Simulation, the variances of Subgrid-Scale terms in the filtered momentum equations are computed both for the dense and perfect gas cases and significant differences are revealed between the two cases. A model using only filtered fields as input is derived for one of the dense gas specific Subgrid-Scale terms using a Correlation Patterns Modeling approach. Preliminary *a priori* tests indicate that this approach can be easily implemented and provides promising results.

1 Introduction

Subgrid terms analysis and modeling for Large Eddy Simulation (LES) in Perfect Gas (PG) flows have been the subject of numerous studies since the pioneering work of Smagorinsky [1] in 1963. A comprehensive review of the different approaches used for LES modeling in compressible flows is available in Garnier *et al.* [2]. In a context where computational power still follows Moore's law, LES now appears as a very important tool for the design of complex systems in the industry, and especially in the Organic Rankine Cycle (ORC) community. On the one hand, only a few research teams have devoted their time and efforts to analyze and model sub-grid terms in the Real Gas (RG) flow context [3]. On the other hand, it has been consistently shown that turbulence is responsible for a significant amount of the losses in dense gas flows [4]. In 2018, a Young Researcher Project has been granted to our group at LMFA by Agence Nationale de la Recherche (ANR) to analyze the fundamental behavior of compressible turbulence in dense gas (DG) flows and to transfer this knowledge to the field of ORC turbine flow modeling. Since then, Direct Numerical Simulation (DNS) has been used to build an extensive database of turbulent DG flows. The database

M. Pini et al. (Eds.): NICFD 2020, ERCOFTAC Series 28, pp. 71–77, 2021.
https://doi.org/10.1007/978-3-030-69306-0_8

comprises decaying Homogeneous Isotropic Turbulence (HIT) [5], forced HIT [6], mixing layer [7] and supersonic channel flow. Using this database and following Vreman's [8] methodology, we analyze in this article the implications of the LES filtering process on the Navier-Stokes equations and the subsequent derivation of sub-grid terms in the DG flow context. As a first step, the evolutions of sub-grid scale (SGS) terms variances with the filter size in the forced HIT are analyzed in Sect. 2 and compared with their PG counterparts. A new modeling approach named Correlation Patterns Modeling is presented in Sect. 3 taking advantage of predictive analytics and machine learning.

2 Analysis of SGS Terms in Dense Gas Flows

2.1 LES: The Issue of Total Energy

When attempting to compute a turbulent compressible flow using Large Eddy Simulation (LES), one has to choose the set of variables to be processed. Garnier *et al.* [2] summarize the various available options, which can be split into two main families, both first described by Vreman [8].

On the one hand, an equation can be derived for the "computable" total energy $\widehat{E}$. This total energy is obtained by using only filtered quantities available from the LES simulation. In perfect gas, with constant heat capacities, it is possible to write an exact computable transport equation for $\overline{\rho}\widehat{E} = \frac{\overline{p}}{\gamma-1} + \frac{1}{2}\overline{\rho}\widetilde{u}_i\widetilde{u}_i$ (with $\overline{\phi}$ and $\widetilde{\phi}$ respectively the filtered and Favre-filtered quantity for a DNS flow field ϕ). By "computable", it is meant the transport equation involves only those filtered quantities provided by the LES simulation. Yet, when the Equation of State (EoS) is more complex, as occurs when describing dense gas flows, it is no longer possible to associate the filtered internal energy to $\frac{\overline{p}}{\gamma-1}$. In that case, non-linearities in the EoS relations prevent the derivation of an exact computable transport equation for the "computable" total energy.

On the other hand, one can maintain in the solver the computation of a transport equation for the filtered total energy $\widetilde{E}$. In this case though, the temperature, often obtained through an iterative process from the knowledge of $\widetilde{E}, \overline{\rho}$ and $\widetilde{u}_i$ is not the Favre-filtered temperature $\widetilde{T}$. It is denoted $\widehat{T}$ and is used with $\overline{\rho}$ in the EoS to compute $\widehat{p}$. This implies that the computed pressure is also not equal to its filtered counterpart $\overline{p}$ appearing in the momentum equation. It also implies that the transport properties needed in the LES, such as $\widehat{\lambda}$ and $\widehat{\mu}$ (dynamic viscosity and thermal conductivity), are not respectively equal to $\overline{\lambda}$ and $\overline{\mu}$ nor for that matter to any simple combination of $\overline{\lambda}$, $\overline{\mu}$, $\widetilde{\lambda}$ or $\widetilde{\mu}$.

Using this approach in the LES context leads to the following equations already derived by Selle *et al.* [3] in the context of multi-species supercritical turbulent flows:

$$\frac{\partial\overline{\rho}}{\partial t} = -\frac{\partial\overline{\rho}\widetilde{u}_j}{\partial x_j} \tag{1}$$

$$\frac{\partial \overline{\rho}\widetilde{u_i}}{\partial t} = -\frac{\partial \widehat{p}}{\partial x_i} - \frac{\partial(\overline{p} - \widehat{p})}{\partial x_i} + \frac{\partial \widehat{\tau_{ij}}}{\partial x_j} + \frac{\partial(\overline{\tau_{ij}} - \widehat{\tau_{ij}})}{\partial x_j} - \frac{\partial \overline{\rho}\widetilde{u_i}\widetilde{u_j}}{\partial x_j} - \frac{\partial \overline{\rho}(\widetilde{u_i u_j} - \widetilde{u_i}\widetilde{u_j})}{\partial x_j} \tag{2}$$

$$\frac{\partial \overline{\rho}\widetilde{E}}{\partial t} = -\frac{\partial \widehat{p}\widetilde{u_i}}{\partial x_i} - \frac{\partial(\overline{pu_i} - \widehat{p}\widetilde{u_i})}{\partial x_i} + \frac{\partial \widehat{\tau_{ij}}\widetilde{u_i}}{\partial x_j} + \frac{\partial(\overline{\tau_{ij}u_i} - \widehat{\tau_{ij}}\widetilde{u_i})}{\partial x_j} - \frac{\partial \widehat{q_i}}{\partial x_i} - \frac{\partial(\overline{q_i} - \widehat{q_i})}{\partial x_j} - \frac{\partial \overline{\rho}\widetilde{E_{int}}\widetilde{u_j}}{\partial x_j} - \frac{\partial \overline{\rho}(\widetilde{E_{int}u_j} - \widetilde{E_{int}}\widetilde{u_j})}{\partial x_j} - \frac{\partial \overline{\rho}\widetilde{K}\widetilde{u_j}}{\partial x_j} - \frac{\partial \overline{\rho}(\widetilde{Ku_j} - \widetilde{K}\widetilde{u_j})}{\partial x_j} \tag{3}$$

where $\widehat{\tau_{ij}} = \widehat{\mu}\left(\frac{\partial \widetilde{u_i}}{\partial x_j} + \frac{\partial \widetilde{u_j}}{\partial x_i} - \frac{2}{3}\frac{\partial \widetilde{u_k}}{\partial x_k}\delta_{ij}\right)$, $\widehat{q_i} = -\widehat{\lambda}\frac{\partial \widehat{T}}{\partial x_i}$ and $K = \frac{1}{2}u_i u_i$.
Written in this way, each term computed using the available filtered variables is associated to its subgrid-scale counterpart arising from correlations at the subgrid level. Two types of SGS terms appear in the set of governing Eqs. (1)–(3):

- Terms $-\frac{\partial \overline{\rho}(\widetilde{u_i u_j} - \widetilde{u_i}\widetilde{u_j})}{\partial x_j}$ in (2) and $-\frac{\partial \overline{\rho}(\widetilde{Ku_j} - \widetilde{K}\widetilde{u_j})}{\partial x_j}$ in (3) can be considered as classical SGS closure terms since they account for unresolved SGS correlations between the different velocity components or between velocity and kinetic energy *regardless of the EoS*. These terms are often the main objective of LES modeling studies.
- The second type of terms comprises both turbulence and thermodynamic information. Their amplitude therefore strongly depends on the complex correlations at small scales between pressure, temperature and mass density.

2.2 Ranking of Momentum SGS Terms Using DNS and Explicit Filtering in a Forced HIT

A forced HIT (with 675^3 elements resolution) is used to compare the RMS amplitudes of all the terms in Eqs. (1)–(3) as a function of the EoS: Perfect Gas (PG) or, for the dense gas (DG), Martin-Hou EoS. The reader is refered to Giauque *et al.* [6] for a complete description of the test case. The explicit filtering size is varied from the size of the computational box ($k/k_{min} = 1$) down to the Nyquist limit ($k/k_{min} = 337$). When the explicit filtering size is equal to the cell size, the subgrid-scale terms tend by definition to zero. For the momentum equations, all terms are normalized by the amplitude of the largest term which is the RMS amplitude of the convective term ($\frac{\partial \overline{\rho}\widetilde{u_i}\widetilde{u_j}}{\partial x_j}$) at the Nyquist limit.

Figure 1 displays the RMS amplitudes of the normalized filtered momentum equations terms. All three vector components of the momentum equation collapse nicely regardless of the EoS, thereby confirming the isotropic nature of the flow. Between the two EoS, the striking difference is the appearance of a new significant term in the DG case: the RMS amplitude of $-\frac{\partial(\overline{pu_i}-\widehat{p}\widetilde{u}_i)}{\partial x_i}$ reaches 25% of the largest term for scales k in the range $[10k_{min}, 100k_{min}]$. Another difference is the decrease of the sub-grid scale term $-\frac{\partial\overline{\rho}(\widetilde{u_iu_j}-\widetilde{u}_i\widetilde{u}_j)}{\partial x_j}$ in the dense gas case from 40% at $k = 10k_{min}$ to 10% at $k = 100k_{min}$. In PG, it remains almost constant and equal to 40% in the same wavenumber range. Note also that for PG the term $\frac{\partial(\overline{\tau_{ij}}-\widehat{\tau_{ij}})}{\partial x_j}$ remains of the same order as the term $\frac{\partial\widehat{\tau_{ij}}}{\partial x_j}$ up to $k = 80k_{min}$.

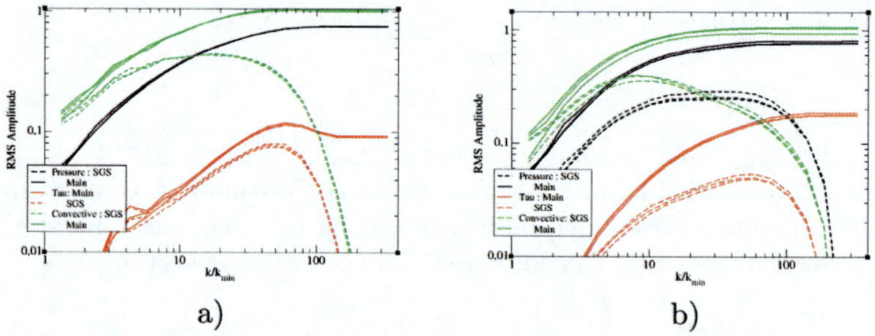

Fig. 1. RMS amplitudes of terms in Eq. (2): a) PG, b) DG

3 Modeling Strategies for SGS Terms in Dense Gas Flows

Several modeling strategies are currently explored for the modeling of the sub-grid terms identified in Eqs. (1)–(3): Artificial Neural Networks (ANN), Random Forest (RF) and Correlation Patterns Modeling (CPM). CPM is inspired by the work of Vollant [9] who used the notion of optimal estimator [10] to develop a sub-grid scale model for the scalar flux in the PG context.

3.1 Correlation Patterns Modeling Applied to DG Flows

As a proof of concept, we consider the modeling of the sub-grid term $sgs = (\widetilde{p}-\hat{p})$ where $\widetilde{p}$ is the filtered pressure and $\hat{p}$ is the pressure obtained from the filtered variables $\widetilde{\rho}, \widetilde{\rho u_i}, \widetilde{\rho E}$. The test case is a 128^3 cells forced HIT solved with the AVBP solver. A single snapshot is considered at a time for which the turbulence is statistically stable. This snapshot therefore represents more than 2 million grid points and constitutes the database to be used for modeling. At each grid point, the following set of flow variables $\{\Phi\}$ is filtered at $k = 10k_{min}$:

$$\{\widetilde{\Phi}\} = \{\widetilde{\rho}, \widetilde{p}, \widetilde{T}, \widetilde{\rho E}, \delta x\widetilde{\nabla(\rho)}, \delta x\widetilde{\nabla(p)}, \delta x\widetilde{\nabla(T)}, \delta x\widetilde{\nabla(\rho E)},$$
$$(\delta x)^2\widetilde{\Delta(\rho)}, (\delta x)^2\widetilde{\Delta(\rho E)}, \delta x\widetilde{I_1}, (\delta x)^2\widetilde{I_2}, (\delta x)^3\widetilde{I_3}\}$$

where I_1, I_2, I_3 are the invariants of the velocity gradient tensor.
Without direct knowledge of sgs, the Optimal Estimator (OE) for this term is defined as $OE =< sgs|\widetilde{\Phi} >$, that is the expectation of sgs conditioned with the set of parameters $\widetilde{\Phi}$. Computing directly this OE would mean discretizing each variable of $\{\widetilde{\Phi}\}$ in 100 bins for instance, to ensure an accurate calculation of the conditional expectation, which would then translate into 100^{13} distinct states in the phase space defined by $\{\widetilde{\Phi}\}$, making this direct approach obviously intractable. Instead, in the present work, we concentrate on conditional expectations limited to 2 variables, $< sgs|\{\widetilde{\Phi_1}, \widetilde{\Phi_2}\} >$ where $\widetilde{\Phi_1}, \widetilde{\Phi_2}$ are a sub-set of $\{\widetilde{\Phi}\}$. Each variable being split into 100 bins, the discrete two variables conditional means are associated to 10^4 states $\{i, j\}$. For each state, along with the conditional average $< sgs|\{\widetilde{\Phi_1(i)}, \widetilde{\Phi_2(j)}\} >$, the variance $rms(< sgs|\{\widetilde{\Phi_1(i)}, \widetilde{\Phi_2(j)}\} >)$ is also computed. Because $\{\widetilde{\Phi}\}$ is composed of 13 variables, 78 separate combinations of two variables $\widetilde{\Phi_1}$ and $\widetilde{\Phi_2}$ can be built to compute averages and variances. The essential idea behind CPM is that if there exists regions of the state space $\{i, j\}$ where

$$| < sgs|\{\widetilde{\Phi_1(i)}, \widetilde{\Phi_2(j)}\} > | \ll rms(< sgs|\widetilde{\Phi_1(i)}, \widetilde{\Phi_2(j)} >)$$

then one can safely use $< sgs|\{\widetilde{\Phi_1(i)}, \widetilde{\Phi_2(j)}\} >$ as a proxy or surrogate for sgs in this region.

The following algorithm is applied to test this idea:

- Compute for each grid point the exact value of sgs
- Compute for all 78 combinations of two variables in $\{\widetilde{\Phi}\}$ the values of $< sgs|\{\widetilde{\Phi_1(i)}, \widetilde{\Phi_2(j)}\} >$ and $rms(< sgs|\{\widetilde{\Phi_1(i)}, \widetilde{\Phi_2(j)}\} >)$ for $i \in [1, 100]$ and $j \in [1, 100]$
- For each $\{i, j\}$, select the combinations satisfying

$$| < sgs|\{\widetilde{\Phi_1(i)}, \widetilde{\Phi_2(j)}\} > | > rms(< sgs|\{\widetilde{\Phi_1(i)}, \widetilde{\Phi_2(j)}\} >)$$

- In that selection, keep the single combination yielding the lowest ratio $\frac{rms(<sgs|\{\widetilde{\Phi_1(i)}, \widetilde{\Phi_2(j)}\}>)}{|<sgs|\{\widetilde{\Phi_1(i)}, \widetilde{\Phi_2(j)}\}>|}$.
- Using this combination and assuming a Gaussian distribution, compute the modeled value as

$$sgs_{\text{predicted}} = \sqrt{rms(< sgs|\{\widetilde{\Phi_1(i)}, \widetilde{\Phi_2(j)}\} >) * random(0, 1)}$$
$$+ < sgs|\{\widetilde{\Phi_1(i)}, \widetilde{\Phi_2(j)}\} >$$

Figure 2 compares the modeled term $sgs_{\text{predicted}}$ to its exact counterpart sgs, with black dots corresponding to a perfect correlation between the model and the exact value. The observed trend for this preliminary assessment indicates the modeling strategy deserves further investigation.

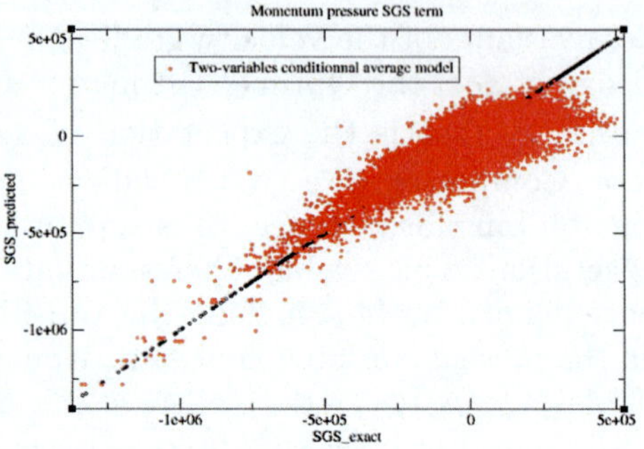

Fig. 2. Distribution of $sgs_{\text{predicted}}$ as a function of sgs

4 Concluding Remarks

In order to develop subgrid-scale (SGS) turbulence models for dense gas flows, SGS terms are first rigorously derived from the filtered Navier-Stokes equations. As already noticed by previous authors, additional terms to the well-known SGS closure terms are identified. Using a reference DNS for dense gas HIT, the variances of SGS terms in the filtered momentum equations are computed both for the dense and perfect gas cases. Significant differences are revealed between DG and PG and it is observed that the non-linearities of the EoS significantly influence the behavior of SGS terms. A surrogate is derived for one of the DG-specific SGS terms using a Correlation Patterns Modeling (CPM) approach. Preliminary *a priori* tests indicate that CPM can be easily implemented and provides promising results, calling for further investigation.

Acknowledgements. This work is supported by the JCJC ANR EDGES project, grant #ANR-17-CE06-0014-01 of the French Agence Nationale de la Recherche. Simulations have been carried out using HPC ressources at CINES under the project grants #A0062A07564 and #A0082A07564.

References

1. Smagorinsky, J.: General circulation experiments with the primitive equations: I. the basic experiment. Monthly Weather Rev. **91**(3), 99–164 (1963)
2. Garnier, E., Adams, N., Sagaut, P.: Large eddy simulation for compressible flows. Springer Science & Business Media (2009)

3. Selle, L., Okong'o, N., Bellan, J., Harstad, K.: Modelling of subgrid-scale phenomena in supercritical transitional mixing layers: an a priori study. J. Fluid Mech. **593**, 57–91 (2007)
4. Durá Galiana, F.J., Wheeler, A.P.S., Ong, J.: A study of trailing-edge losses in organic rankine cycle turbines. J. Turbomach., **138**(12), 121003 (2016)
5. Giauque, A., Corre, C., Menghetti, M.: Direct numerical simulations of homogeneous isotropic turbulence in a dense gas. J. Phys. Conf. Ser. **821**(1), 12017 (2017)
6. Giauque, A., Corre, C., Vadrot, A.: Direct numerical simulations of forced homogeneous isotropic turbulence in a dense gas. J. Turbul. **21**(3), 186–208 (2020). https://doi.org/10.1080/14685248.2020.1742916
7. Vadrot, A., Giauque, A., Corre, C.: Analysis of turbulence characteristics in a temporal dense gas compressible mixing layer using direct numerical simulation. J. Fluid Mech. **893**, A10 (2020)
8. Vreman, B., Geurts, B., Kuerten, H.: Subgrid-modelling in LES of compressible flow. Appl. Sci. Res. **54**(3), 191–203 (1995)
9. Vollant, A., Balarac, G., Corre, C.: Subgrid-scale scalar flux modelling based on optimal estimation theory and machine-learning procedures. J. Turbul. **18**(9), 854–878 (2017)
10. Balarac, G., Pitsch, H., Raman, V.: Development of a dynamic model for the subfilter scalar variance using the concept of optimal estimators. Phys. Fluids **20**(3), 035114 (2008)

Tolerance Optimization of Supersonic ORC Turbine Stator

Nassim Razaaly[1(✉)], Giacomo Persico[2], and Pietro Marco Congedo[1]

[1] Inria Saclay IDF, Palaiseau, France
{nassim.razaaly,pietro.congedo}@inria.fr
[2] Politecnico di Milano, Milan, Italy
giacomo.persico@polimi.it

Abstract. The discrepancy between manufactured and design geometry of turbomachinery blades has a detrimental effect on the performance variability. In this work, the authors propose a methodology to reduce the impact of the randomness induced by the manufacturing process: a *tolerance* optimization is carried out by resorting to an efficient robust optimization method based on quantile regression. Its application to a typical two-dimensional supersonic nozzle cascade for ORC showcases promising preliminary results.

Keywords: ORC Stator · Quantile regression · NICFD · Robust optimization · Random field · Geometric uncertainty

1 Introduction

The geometry of manufactured turbomachinery blades inevitably differs from the design geometry, due to noise in the manufacturing process or in-service erosion. It generally induces both an increase in performance variability while decreasing its mean performance. Garzon [5] demonstrated that the mean loss coefficient of a flank-milled integrally bladed rotor (IBR) increased by 23% due to manufacturing variability. Dow proposed a robust design framework combining both geometric and tolerance optimization of a compressor exit vane, leading to a quite large number of flow simulations performed using the Multiple Blade Interacting Streamtube Euler Solver (MISES) code [3]. Similarly to Dow, we assume that the geometric variability in manufactured turbine blades can be described as a non-stationary Gaussian Random Field, representing the error between the manufactured surface and the nominal (perfect) one, fully defined by its autocovariance function.

The effect of tightening manufacturing tolerances is modelled by reducing the standard deviation of the random field locally. This work focuses on the tolerance optimization by resorting to a Robust Optimization (RO) based on Quantile Regression [14], yielding a low number of CFD simulations. The proposed method permits to quantitatively highlight the regions of the blade that have the largest impact on the average performance.

This methodology is applied to a typical supersonic nozzle cascade for ORC applications, using the popular NICFD flow solver SU2. Indeed, the peculiarities of organic

M. Pini et al. (Eds.): NICFD 2020, ERCOFTAC Series 28, pp. 78–86, 2021.
https://doi.org/10.1007/978-3-030-69306-0_9

fluids typically lead to supersonic turbine configurations featuring supersonic flows and shocks, which can be influenced by the geometric tolerances of the blade manufacturing, as emphasized in [11].

The supersonic nature of the blade makes the throat zone, especially in the supersonic area downstream the throat (both in the divergent and in the semi-bladed part), very *sensitive* to the geometry. This behavior might be exploited in the manufacturing phase, only if the effect is appropriately modeled and quantified. This long term objective motivates the present research work.

In this paper, we propose an efficient framework for the robust design of the tolerances of a two-dimensional nozzle cascade. The originality w.r.t. [3] lies in the use of a parcimonious RO method combined with an original parametrization of the design space.

2 Methodology

The NICFD test-case is described in Subsect. 2.1. The uncertainty framework and in particular the description of the perturbed blades generation from a given design vector is fully presented in Subsect. 2.2. The robust optimization method is described in Subsect. 2.3.

2.1 NICFD Case Description

The targeted Organic Rankine Cycle (ORC) turbine is the geometry of an existing ORC stator designed for a 300 kWe Combined-Heat-and-Power axial turbogenerator employing siloxane MDM as working fluid (whose properties are reported in Table 1). This supersonic axial-flow turbine stator is characterized by converging-diverging blades and it features significant fluid-dynamic penalties due to a strong shock-wave forming on the rear suction side of the blade. This exemplary profile, originally presented in [1] has been extensively studied in the open literature of ORC. In recent years, it has been subjected to several deterministic optimization trials [9, 16] and robust ones in [10, 12]. The flow model focuses on the two-dimensional flow around the blade profiles at the midspan section of the cascade considering operating conditions provided in Table 1.

In order to estimate the aerodynamic performances of the supersonic turbine, the Non-Ideal Compressible-Fluid Dynamics solver included in the SU2 [4, 8, 17] suite is employed, embedding in particular the Peng-Robinson-Stryjek-Vera Equation of State to describe the fluid thermodynamic behavior. Inviscid fluxes are discretized using a MUSCL approach based on an approximate Riemann solver of Roe upwind type [7, 13, 15] along with the slope limiter proposed by van Albada. Additionally, Non-Reflecting Boundary Conditions [6] are exploited to avoid spurious pressure oscillations due to the reflection of spurious pressure waves at domain boundaries. The unstructured grids are generated using an in-house tool based on an advancing-front/Delaunay algorithm. Inviscid simulations are performed using a restart file corresponding to the baseline simulation, on meshes constituted of 16k cells, selected in [12] after grid dependence study, as optimal trade-off between accuracy and cost (Table 2).

Table 1. MDM gas properties and operating conditions.

Critical pressure	14.152 bar
Critical temperature	564.1 K
Critical density	256.82 kg.m^{-3}
γ	1.0165
Acentric factor ω	0.529
Gas constant	35.152 J/kg/K
μ	1.1517×10^{-5} Pa.s
k	0.03799 W/(m.K)
Inlet total pressure [Pa]	8×10^5
Inlet total temperature [K]	545.15
Inlet axial angle [°]	0
Outlet static pressure [Pa]	1.072×10^5

Table 2. Test-case: mach contours and 16k cells mesh.

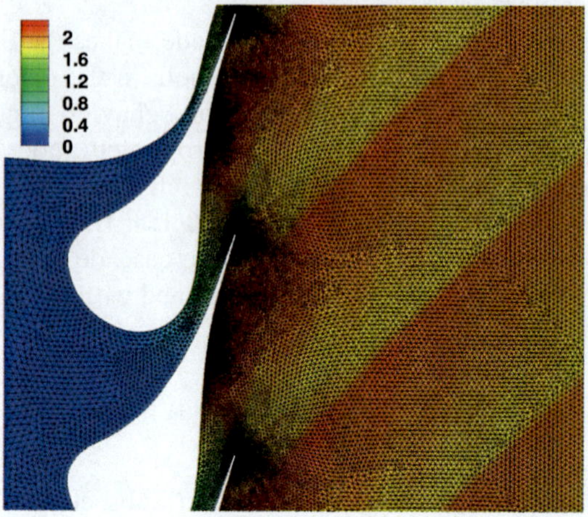

The Quantity of Interest (QoI) considered is the standard deviation of the distribution of the static pressure measured half-axial chord downstream the blade, denoted as ΔP. This objective function is selected since minimizing it directly increases the performance of the stator as it minimizes the impact of the shock, which is the main source of loss here. It also has a beneficial effect on the downstream rotor, as it makes the flow entering it more uniform. Figure 2 illustrates the NICFD test-case by providing the Mach contours and showcasing the mesh.

2.2 Modeling Geometric Variability

2.2.1 Random Field Generation

Following [2,3,11], we assume that the geometric variability in manufactured turbine blades can be accurately described as a *non-stationary Gaussian Random Field* $e(s,\omega)$, ω being a coordinate in the sample space Ω, and $(\Omega, \mathscr{F}, \mathbb{P})$ a complete probability space. The arclength $s \in [0,1]$ parametrizes the location on the blade surface, starting at the trailing edge ($s=0$), going around the leading edge ($s=\frac{1}{2}$), and continuing back to the trailing edge on the opposite side of the blade ($s=1$).

The Random field $e(s,\omega)$ represents the error between the manufactured surface and the nominal (perfect) one in the normal direction at the point parametrized by s. It is fully defined by its mean $\bar{e}(s)$ (null here) and autocovariance function $C(s,t)$: it captures the correlation between manufacturing errors at locations along the blade surface parametrized by $s, t \in [0,1]$, and describes the smoothness and correlation length of the random field. It is written as [11]:

$$C(s,t) = \sigma(s)\ \sigma(t)\ \rho(s,t) \tag{1}$$

where $\sigma(s)$ is the standard deviation of the random field at location s, which quantifies locally the level of manufacturing variability. Its modeling is described in Subsect. 2.2.2, and is the goal of the present work. The non-stationary autocorrelation function ρ is defined by:

$$\rho(s,t) = \exp\left(-\frac{|s-t|^2}{L(s)L(t)}\right) \tag{2}$$

where

$$L(s) = L_0 + (L_{LE} - L_0)\exp\left(-\frac{|s-\frac{1}{2}|^2}{w^2}\right) \tag{3}$$

The values $L_0 = 0.1$, $w = 0.1$ and $L_{LE} = 1.0 \times 10^{-2}$, all normalized by the blade half-arclength were used [11]. In the present study, the trailing edge is modeled as a straight segment, thus, the impact of manufacturing variability at the trailing edge is not addressed.

In order to sample a random field $e(\cdot,\omega)$ on a set $\mathscr{S} = \{s_1,...,s_{n_{blade}}\}$, representing the values in $[0,1]$ parametrizing the n_{blade} nodes on the blade profile, the following procedure is applied:

- The discrete covariance matrix $\mathbf{C} = [C(s_i,s_j)]_{i\in[\![1,n_{blade}]\!]}$ is generated.
- A gaussian vector $g \in \mathbb{R}^{n_{blade}}$ is generated, with $g \sim \mathscr{N}(0,\mathbf{C})$.

For more details, the reader is referred to [11].

2.2.2 $\sigma(s)$ Parametrization: Design Vector

A standard approach [11] consists in assuming a constant standard deviation of the random field: $\sigma(s) = \sigma_0$, for a given constant $\sigma_0 = 3 \times 10^{-5}m$ here. The goal of the present work is to find an optimal distribution of $\sigma(\cdot)$ leading to a reduced detrimental impact of geometric variability on the prescribed blade profile, satisfying a (manufacturing) cost constraint. Here we simply search for a distribution satisfying $\int_{[0,1]} \sigma(s)ds = \sigma_0$. A two-degree of freedom parametrization is chosen here, based:

$$\sigma_{\mathbf{x}}(s) = \sigma_0 + b(s,\mathbf{x}) \tag{4}$$

where $b(s,\mathbf{x})$ is a cubic Béziers curve in $[0,1]$, defined by 4 so-called control points $P_0,P_1,P_2,P_3 \in \mathbb{R}^2$:

$$b(s,\mathbf{x}) = P_0(1-s)^3 + 3P_1 s(1-s)^2 + 3P_2 s^2(1-s) + P_3 s^3. \tag{5}$$

To ensure $\int_{[0,1]} (s,\mathbf{x})ds = 0$, we choose $P_0 = (0,x_2)$, $P_1 = (0.5 - x_0, x_1)$, $P_2 = (0.5 + x_0, -x_1)$ and $P_3 = (0,-x_2)$, with $\mathbf{x} = (x_0,x_1,x_2)$ denoting the design vector. The design space is chosen as $\Omega = [-3,3] \times [-2\sigma_0, 2\sigma_0] \times [\sigma_0, \sigma_0]$. Figure 1 provides an illustration of this parametrization, and the corresponding influence of the local level of manufacturing variability on the blade. In the following, a random field as described in Subsect. 2.2.1 based on the standard deviation $\sigma_{\mathbf{x}}(s)$ is denoted by $e_{\mathbf{x}}(s,\omega)$.

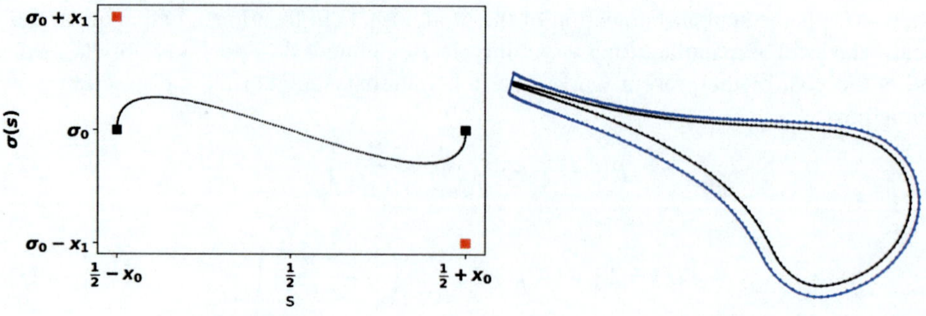

(a) 1D Representation (b) Blade

Fig. 1. Random Field Illustration. (a) Distribution $\sigma_{\mathbf{x}}(s)$, with $\mathbf{x} = (0.5, 2\sigma_0, 0)$. (b) Blade geometry. Baseline (black) and corresponding $\sigma_{\mathbf{x}}(s)$ perturbation (blue) with scale = 50.

Figure 2 presents examples of realizations for a given design vector.

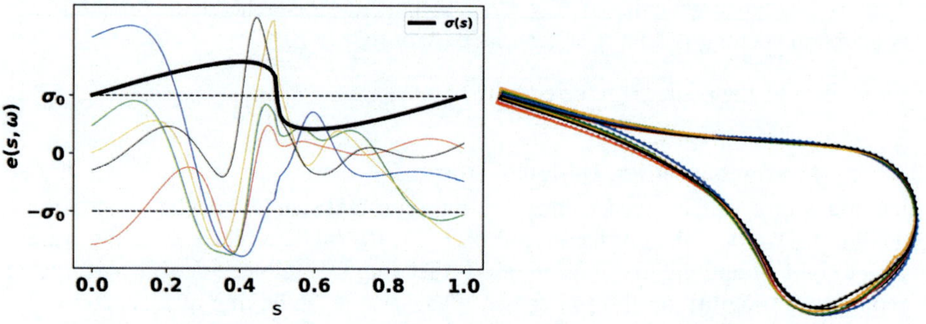

(a) 1D Representation (b) Blade

Fig. 2. Random field realizations: $\mathbf{x} = (0.5, 2\sigma_0, 0)$, scale = 20.

2.2.3 Stochastic Model

For a given design vector $\mathbf{x}$, a random field $e_{\mathbf{x}}$ is defined based on its local standard deviation distribution $\sigma_{\mathbf{x}}(\cdot)$. A geometric error realization $e_{\mathbf{x}}(\cdot, \omega^*(\mathbf{x})$ following $e_{\mathbf{x}}$ is then sampled, corresponding to a perturbed blade geometry, where $\omega^*(\mathbf{x})$ representing a coordinate in the sample space depending on $\mathbf{x}$. A CFD simulation permits the evaluation of the QoI $\Delta P(\mathbf{x}, \omega^*(\mathbf{x})$, or simply the random QoI $\Delta P(\mathbf{x})$. In other terms, $\mathbf{x} \mapsto \Delta P(\mathbf{x})$ can be represented as a scalar stochastic model, as several simulations of the same input normally result in different outputs.

2.3 (Robust) Optimization Framework

The objective of the present work is to find an optimal distribution of the standard deviation of the random field describing the geometric discrepancy between manufactured blades and the nominal (perfect) one. Such optimal distribution would permit to limit the detrimental effect on the variability of the scalar random QoI ΔP. We propose to recast this problem as a robust optimization one, formulated as follows:

$$\begin{aligned} \text{Minimize} \quad & q_{80}[\Delta P(\mathbf{x})] \\ \text{s.t.} \quad & \mathbf{x} \in \Omega \subset \mathbb{R}^3 \end{aligned} \tag{6}$$

where q_{80} is the 80% quantile operator induced by the random field $e_{\mathbf{x}}$. Other statistics could have been chosen, such as linear combination of the mean and standard deviation, other quantiles... In order to solve Eq. 6, we resort to a methodology based on Quantile Regression (QR) [14] for Robust Optimization. This method is suitable for low dimensional design space, and high dimensional (possibly non-parametric) stochastic space, which is consistent with the problem we propose to solve. An initial Design of Experiment (DoE) of size $n0 = 300$ is first generated $\mathscr{T} = \{\mathbf{x}_i, \Delta P(\mathbf{x}_i)\}_i$, based on Latin Hypercube Sampling (LHS). A QR based on $n_{CP} = 37$ so-called Control Points is built. The seminal work described in [14] is slightly modified: the bayesian description of the QR enabling to obtain uncertainties in the QR derivation based on so-called Markov-Chain Monte-Carlo (MCMC) is replaced by building several QR representations based on random generation of the Control Points. It enables to cheaply obtain an estimation of the mean and quantiles of the QR, used to select promising designs. At each optimization iteration, $K = 8$ CFD runs are performed to update the DoE.

3 Results

Some preliminary results are presented: the optimal configuration is compared w.r.t. the baseline one $\mathbf{x}_0$, which corresponds to a constant standard deviation distribution, *i. e.* $\sigma(s) = \sigma_0$. The robust optimization framework required 38 iterations to converge, for which 8 CFD simulations are run in parallel in less than 2 m, so that the final number of CFD run is $nf = 604$. To assess the quantile results for both the baseline $\mathbf{x}_0$ and the optimal design $\mathbf{x}^*$, 200 samples following respectively $e_{\mathbf{x}_0}$ and $e_{\mathbf{x}^*}$ are generated and assessed by the NICFD solver. First, the so-called Probability Density Function (PDF) of the QoI ΔP is built, as the empiric 80% quantile, as reproduced in Fig. 3(a). We can observe that the PDF corresponding to the optimized design is significantly less spread than for the baseline configuration, while the difference between both quantiles is small (0.15%). This noticeable favorable impact may be due to the choice of a high quantile for the statistics. Same comments apply to the total loss pressure coefficient Y PDF 3(b).

Figure 5 represents the local distribution of the standard deviation for the optimized case, and examples of realizations of perturbed blade geometries.

To exhibit the impact of the distribution of the standard deviation on the blade aerodynamics, the contours of Coefficient of Variation (CoV) of the Mach number is presented in Fig. 4, with $Cov = \dfrac{\sigma}{\mu}$. The CoV values are estimated locally at the nodes of the mesh, using standard empirical estimators. The Mach CoV for the baseline in Fig. 4(a) is very similar to the analysis obtained in [11] from a RANS simulation. The result obtained for the optimized design emphasizes that the CoV increases in the shock zone, while it decreases in the isentropic region upstream the blade and at the beginning of the bladed channel.

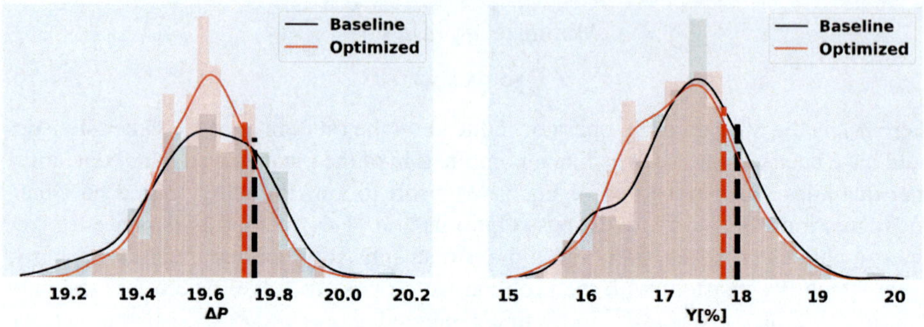

(a) ΔP PDF (b) Y PDF

Fig. 3. PDF of the QOIs ΔP (a) and total pressure loss Y (d): baseline design $\mathbf{x}_0$ (black) and optimal design $\mathbf{x}^*$ (red). The vertical line exhibits the 80% quantile.

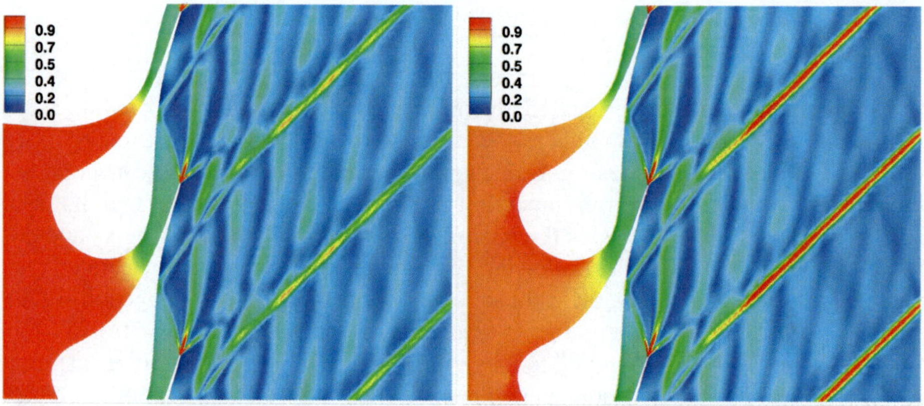

(a) Baseline (b) Optimized

Fig. 4. Contours of mach CoV[%] for the baseline and the optimal designs, based on 200 CFD samples.

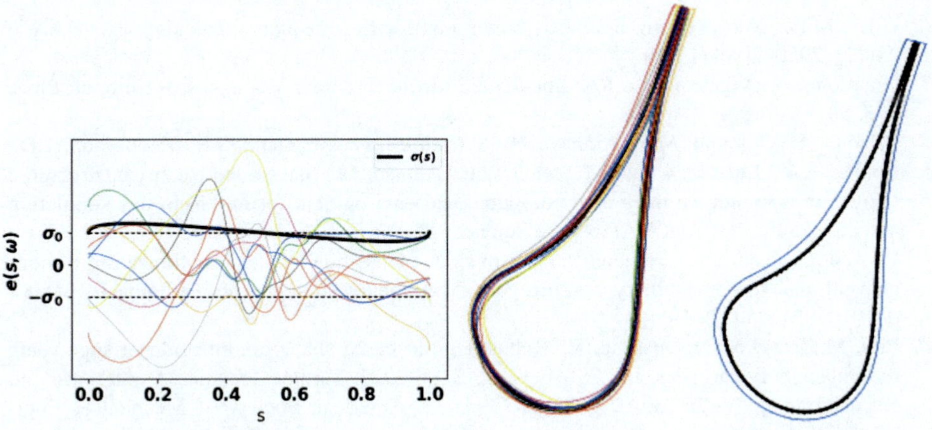

(a) 1D Representation (b) Blade (c) Standard Deviation

Fig. 5. Optimized design: random field realizations and local distribution of the standard deviation (scale = 20).

4 Conclusions and Future Works

The present research work proposed a methodology for optimizing the distribution of the *geometric variability*. This problem is solved by resorting to a modified state-of-the-art robust optimization method based on QR. We demonstrated its impact applying it to a two-dimensional nozzle cascade, using an inviscid NICFD solver. The robust optimization problem, characterized by a design space of small dimensionality ($d = 3$) and large stochastic dimension, required at total of 608 CFD evaluations: $n_0 = 300$ for the initialization, and 37 iterations of $K = 8$ CFD runs performed in parallel. A significant improvement of the statistics and the PDF of the QOIs is obtained. Those are still preliminary results. Future work will consist mainly in improving the parametrization of the design space and considering RANS simulations.

References

1. Colonna, P., Harinck, J., Rebay, S., Guardone, A.: Real-gas effects in organic Rankine cycle turbine nozzles. J. Propul. Power **24**(2), 282–294 (2008)
2. Dow, E.A., Wang, Q.: Optimal design and tolerancing of compressor blades subject to manufacturing variability. In: 16th AIAA Non-Deterministic Approaches Conference, p. 1008 (2014)
3. Dow, E.A., Wang, Q.: The implications of tolerance optimization on compressor blade design. J. Turbomach. **137**(10) (2015)
4. Economon, T.D., Mudigere, D., Bansal, G., Heinecke, A., Palacios, F., Park, J., Smelyanskiy, M., Alonso, J.J., Dubey, P.: Performance optimizations for scalable implicit RANS calculations with SU2. Comput. Fluids **129**, 146–158 (2016)
5. Garzon, V.E.: Probabilistic aerothermal design of compressor airfoils. PhD thesis, Massachusetts Institute of Technology (2003)

6. Giles, M.B.: Nonreflecting boundary conditions for Euler equation calculations. AIAA J. **28**(12), 2050–2058 (1990)
7. Guardone, A., Vigevano, L.: Roe linearization for the van der Waals gas. J. Comput. Phys. **175**, 50–78 (2002)
8. Palacios, F., Colonno, M.F., Aranake, A.C., Campos, A., Copeland, S.R., Economon, T.D., Lonkar, A.K., Lukaczyk, T.W., Taylor, T.W.R., Alonso, J.J.: Stanford university unstructured (SU2): an open-source integrated computational environment for multi-physics simulation and design. In: 51st AIAA Aerospace Sciences Meeting and Exhibit (2013)
9. Persico, G., Rodriguez-Fernandez, P., Romei, A.: High-fidelity shape optimization of non-conventional turbomachinery by surrogate evolutionary strategies. J. Turbomach. **141**(8), 081010–0810111 (2019)
10. Pini, M., Persico, G., Dossena, V.: Robust adjoint-based shape optimization of supersonic turbomachinery cascades. In: ASME Turbo Expo 2014: Turbine Technical Conference and Exposition, pp. V02BT39A043–V02BT39A043. American Society of Mechanical Engineers (2014)
11. Razaaly, N., Persico, G., Congedo, P.M.: Impact of geometric, operational, and model uncertainties on the non-ideal flow through a supersonic ORC turbine cascade. Energy **169**, 213–227 (2019)
12. Razaaly, N., Persico, G., Gori, G., Congedo, P.M.: Quantile-based robust optimization of a supersonic nozzle for organic Rankine cycle turbines. Appl. Math. Model **82**, 802–824 (2020)
13. Roe, P.L.: Approximate Riemann solvers, parameter vectors, and difference schemes. J. Comput. Phys. **43**(2), 357–372 (1981)
14. Sabater, C., Le Maitre, O., Congedo, P.M., Gortz, S.: A Bayesian approach for quantile optimization problems with high-dimensional uncertainty sources. Researchgate (2020)
15. Vinokur, M., Montagné, J.L.: Generalized flux-vector splitting and roe average for an equilibrium real gas. J. Comput. Phys. **89**, 276 (1990)
16. Vitale, S., Albring, T.A., Pini, M., Gauger, N.R., Colonna, P., et al.: Fully turbulent discrete adjoint solver for non-ideal compressible flow applications. J. Global Power Propul. Soc. **1**, Z1FVOI (2017)
17. Vitale, S., Gori, G., Pini, M., Guardone, A., Economon, T.D., Palacios, F., Alonso, J.J., Colonna, P.: Extension of the SU2 open source CFD code to the simulation of turbulent flows of fuids modelled with complex thermophysical laws. AIAA Paper **2760**, 2015 (2015)

Numerical Investigation of a Supercritical CO_2 Centrifugal Compressor with an In-House Density Based Compressible CFD Solver

Renan Emre Karaefe[1(✉)], Pascal Post[1], Marwick Sembritzky[1], Andreas Schramm[1], Francesca di Mare[1], Matthias Kunick[2], and Uwe Gampe[3]

[1] Ruhr University Bochum, Faculty of Mechanical Engineering, Chair of Thermal Turbomachines and Aeroengines, 44780 Bochum, Germany
{emre.karaefe,pascal.post,marwick.sembritzky,andreas.schramm, francesca.dimare}@rub.de

[2] Zittau/Goerlitz University of Applied Sciences, Faculty of Mechanical Engineering, Department of Energy Systems Technology, 02763 Zittau, Germany
m.kunick@hszg.de

[3] Technische Universität Dresden, Faculty of Mechanical Science and Engineering, Chair of Thermal Power Machinery and Plants, 01062 Dresden, Germany
uwe.gampe@tu-dresden.de

Abstract. This work presents a numerical investigation of the performance characteristics, flow field and internal aerodynamic losses of a supercritical CO_2 centrifugal compressor. The investigated compressor geometry is deduced from main dimensions of a sCO_2 test-loop main compressor operated by Sandia National Laboratories. Three-dimensional CFD simulations are performed with an in-house hybrid CPU/GPU compressible CFD solver. An efficient and accurate real gas property tabulation technique, the Spline-Based Table Look-Up Method (SBTL), which is particularly optimised for the densitiy based solution method, is coupled to the flow solver to account for thermophysical properties. CFD simulations are compared to experimental data as well as a 1D-meanline analysis.

Keywords: sCO_2 · Centrifugal compressor performance · Real gas CFD and property tabulation · 1D-meanline analysis

1 Introduction

Recently, supercritical CO_2 Brayton cycles have gained increasing attention due to their comparatively high thermal efficiency (≈38–50%) at moderate cycle temperatures (450–600 °C) [1]. This is possible because significantly less compression work is required when compressors are operated near the critical point with high fluid density. A high fluid density also allows for a compact turbomachinery design and an overall small footprint of the respective cycle.

M. Pini et al. (Eds.): NICFD 2020, ERCOFTAC Series 28, pp. 87–99, 2021.
https://doi.org/10.1007/978-3-030-69306-0_10

Besides stable process control and compact heat exchanger design, satisfactory turbomachinery performance levels leave room for cycle improvements. Although centrifugal compressor design and analysis is sophisticated for fluids which show ideal gas behaviour, the applicability of respective methods for real fluids showing non-linear behaviour, especially in near-critical state, is uncertain. In this regard, high fidelity CFD simulations, with an accurate thermophysical property representation, allow for detailed insight into the flow field and promise to be an effective tool to improve aerodynamic design and analysis.

The most accurate equation of state for CO_2 is given by Span & Wagner (SW-EOS) [29] and is predominantly integrated into numerical simulations of sCO_2 compressors through lookup tables, due to its high computational overhead. Most noticeably, a sCO_2 main compressor installed in a compression test-loop operated by Sandia National Laboratories (SNL) [32,34] has been investigated through this approach by several researchers. The referenced case is of great interest and also investigated in this work because rare geometrical and experimental data is partially documented. In numerical studies related to the SNL compressor, bilinear interpolation schemes were predominantly applied to compute properties between discrete tabulated values. These require extensive table refinements in order to capture highly non-linear property behaviour. According to Ameli et al. [2,4], such refinements are often the cause for simulation instabilities, which can be coped by a labour-intensive initialisation procedure. In this regard, the applied numerical framework based on the application of biquadratic spline interpolation promises improved simulation stability as well as desirable computing times.

2 Scope of Research

The numerical framework, comprising the in-house CFD solver coupled to the SBTL method, is applied for performance as well as flow field assessments of a centrifugal compressor operating with CO_2. The investigated geometry is deduced from main dimensions of the SNL main compressor, which is designed for operation near the critical point, and has been experimentally investigated by Wright et al. [32,34] and Fuller & Eisemann [12]. Beside a near-critical thermodynamic inlet condition, also a compressor inlet state in the gas phase is considered, where the compressor could potentially be operated during cycle startups. Complementary to the CFD assessments, previously reported in [18], a 1D-meanline analysis is conducted in this study to provide a further reference for performance calculations and to estimate as well as separate the contribution of internal loss mechanisms.

3 Methodology

3.1 Numerical Procedure

The in-house density based compressible CFD solver SharC [25] is applied for all simulations in this study. The Reynolds-averaged Navier-Stokes equations are solved, which are closed by general thermodynamic state expressions as well as constitutive equations. Governing equations are spatially discretised on structured grids using a finite-volume approach. A central scheme is applied for diffusive flux discretisation, while advective fluxes are discretised through a second order AUSM+ scheme [24] using a piecewise linear MUSCL reconstruction [30]. An implicit LUSGS scheme is utilised for time integration and closure of the RANS-equations is achieved by applying the turbulence model of Spalart & Allmaras [28]. Recent solver applications comprise the topics of non-equilibrium wet steam [25], involving the first LES of a condensing wet steam turbine cascade [26], humid air [14] and sCO_2 [18].

3.2 Real Gas Property Tabulation

Real gas thermodynamic properties of CO_2 based on the SW-EOS as well as thermal conductivity and viscosity are accessible in the solver via a software library integration of the Spline-Based Table Look-Up Method (SBTL), specifically developed for fast and accurate fluid-property computations in extensive numerical simulations [19,20]. The method is based on spline interpolation techniques using lower order polynomials, such as biquadratic or bicubic spline functions. Unlike local bilinear interpolation, biquadratic spline functions provide continuous first derivatives. Furthermore, biquadratic spline functions can be solved quickly with respect to their independent variables. In this way, numerically consistent backward functions can be calculated, which are faster than those obtained from the inversion of bicubic polynomials. Also, simple search algorithms are applicable as the construction of spline functions is performed on piecewise equidistant nodes. The permissible deviations are smaller than the uncertainties of the SW-EOS and the transport property correlations, except for thermal conductivity and isobaric heat capacity, which become infinite at the critical point. Throughout the simulations, thermodynamically stable states are considered only, i.e. fluid properties beyond the saturation curves are calculated as a homogenous equilibrium mixture in the two-phase region.

3.3 Geometry Generation and Computational Setup

The impeller geometry of the SNL compressor is investigated, which is not publicly reported in all details. Therefore, some main dimensions given in [34], along with additional part drawings reported in [32,34], were used to design an impeller with comparable main dimensions. For the sake of simplification, tip clearance was not considered. A 3D geometric model (see Fig. 1a) was generated through fixed inputs to a preliminary design-tool and its corresponding interface to a blade generation module [7,8].

The computational domain (see Fig. 1b) was generated with an automated multi-block structured mesh generator [22]. Because of rotational periodicity, a single blading passage comprising approximately 1.7 million cells (max. expansion ratio ≈ 2.6, min. skewness angle $\approx 19°$) and a near-wall clustering was considered. Constant total temperature and total pressure corresponding to

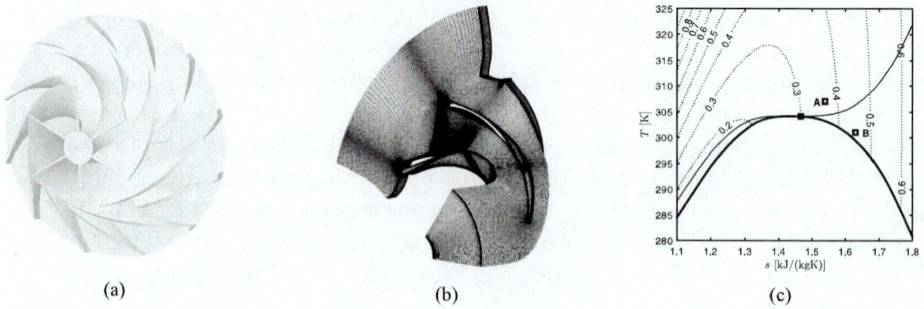

Fig. 1. (a) Impeller wheel geometry, (b) Computational domain, (c) Ts-diagram with markers corresponding to the investigated compressor inlet total states; state A: T_t = 307 K, p_t = 77.5 bar; state B: T_t = 301 K, p_t = 67.9 bar; contour lines show the compressibility factor.

the two investigated inlet states, which are stated and illustrated in Fig. 1c, as well as a turbulent viscosity ratio were imposed at the domain inlet. A constant static pressure was defined as the outlet boundary condition. The imposed outlet static pressure values were (92, 93, 94, 95, 96, 96.5, 96.75) bar to calculate the compressor performance speedline at 50 krpm rotational speed for the near-critical total inlet state (state A). For compressor operation at a total inlet condition in the gas phase (state B), outlet static pressure values of (78, 78.25, 78.5, 79, 80, 81, 81.5, 81.75) bar were specified to derive the 50 krpm performance speedline. The no-slip condition was applied at all walls, which were defined as adiabatic. A routine initialisation procedure was applicable for the RANS calculations, even for simulations with a specified inlet state close to the critical point. The solution domain of the first calculation of each speedline assessment was initialised with spatially linear interpolated values between the domain inlet and outlet for the cyclindrical velocity components as well as pressure and temperature. These were guessed from initial meanline calculations. Consecutive RANS calculations were initialised from the solution of previous converged runs. Compared to ideal gas calculations, which are used as a benchmarking reference, the coupling of the solver to the SBTL library resulted in a computational overhead of merely 33%, providing desirable time-efficiency.

3.4 Meanline Analysis Procedure

In this work, a single-zone modelling approach is applied, which incorporates a loss model set that was suggested by Oh et al. [23] as the most suitable among the possible combinations of most of the loss models in the literature. The loss model set was validated by Oh et al. for a variety of centrifugal compressors operating with working fluids obeying the ideal gas law and also adopted and stated to be valid for the calculation of the SNL sCO_2 main compressor [3,21,27]. For impeller calculations, six sources of internal losses and three sources of external losses are distinguished, which are all expressed in an enthalpy (adiabatic head loss) formulation and listed in Table 1. External losses, which are also called parasitic losses, differ from internal aerodynamic losses in that they give rise to the impeller stagnation enthalpy, but do not contribute to pressure rise:

$$\Delta h_t = W_{Euler} + \sum W_{external} \overset{\text{here}}{=} W_{Euler} + W_{df} + W_{rc} + W_{lk} \quad \text{(assuming adiabatic flow).} \tag{1}$$

In addition to the loss model set, a slip factor correlation

$$\sigma = 1 - \frac{\sqrt{\cos \beta_{2,bl}}}{Z_{bl}^{0.7}} \quad \text{(meridional angle system),} \tag{2}$$

based on a formulation given by Wiesner [31] is implemented, which is necessary in order to deduce the actual work input of the impeller blades through the Euler equation, compared to the case where the fluid would be perfectly guided by the blades (e.g. if there were an infinite number of infinitesimally thin blades):

$$W_{Euler} = u_2 c_{2\theta} - u_1 c_{1\theta} = u_2(\sigma u_2 + c_{2m} \tan \beta_{2,bl}) \quad \text{(meridional angle system).} \tag{3}$$

The meanline analysis procedure is implemented in PYTHON and accounts for the thermophysical properties of sCO_2 by direct calls to the open-source property library CoolProp [6]. In addition, an ideal gas calculation procedure is available, that is suitable for performance calculations of centrifugal compressor impellers operating with conventional working fluids, which obey the ideal gas equation. A sample validation example for performance prediction capabilities of the latter is presented in Fig. 2, based on the well documented NASA LSCC air compressor impeller test case [13].

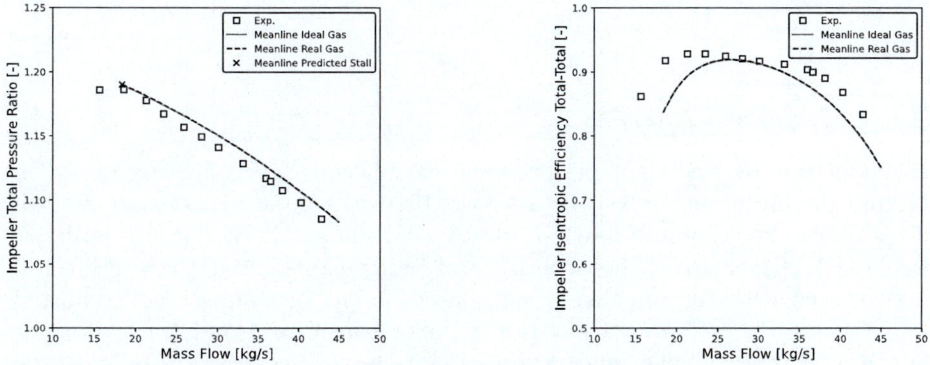

Fig. 2. Meanline performance analysis of the NASA LSCC air centrifugal impeller [13] conducted for design rotational speed on the basis of the real gas and the ideal gas calculation routine. The stall indication criterion is taken from [5].

Table 1. Set of internal and external (parasitic) loss models

Loss mechanism	Loss model		Reference
Incidence	$\Delta h_{inc} = \sigma \cdot \frac{(w_{1\theta} - w_{1\theta,bl})^2}{2}$	$\sigma = 0.5...0.7$	Conrad et al. [9]
Blade loading	$\Delta h_{bl} = 0.05 \cdot D_f^2 \cdot u_2^2$	$D_f = 1 - \frac{w_2}{w_{1t}} + \frac{0.75 \cdot W_{Euler}/u_2^2}{w_{1t}/w_2 \left[\frac{Z_{bl}}{\pi}\left(1 - \frac{d_{1t}}{d_2}\right) + 2\frac{d_{1t}}{d_2}\right]}$	Coppage et al. [10]
Skin friction	$\Delta h_{sf} = 2c_f \frac{L_{fl}}{d_{hb}} \bar{w}^2$	$\bar{w} = \frac{(2w_2 + w_{1t} - w_{1h})}{4}$	Jansen [15]
		$L_{fl} \approx \frac{\pi}{8}\left(d_2 - \frac{d_{1t}+d_{1h}}{2} - b_2 + 2L_{ax}\right)\left(\frac{2}{\frac{\cos\beta_{1t}+\cos\beta_{1h}}{2} + \cos\beta_2}\right)$	
		$d_{hb} = \frac{d_2 \cos\beta_2}{\frac{Z_{bl}}{\pi} + \frac{d_2 \cos\beta_2}{b_2}} + \frac{\frac{d_2}{2}\left(\frac{d_{1t}}{d_2} + \frac{d_{1h}}{d_2}\right)\left(\frac{\cos\beta_{1t}+\cos\beta_{1h}}{2}\right)}{\frac{Z_{bl}}{\pi} + \left(\frac{d_{1t}+d_{1n}}{d_{1t}-d_{1n}}\right)\left(\frac{\cos\beta_{1t}+\cos\beta_{1h}}{2}\right)}$	
		$c_f = 0.0412 Re^{-0.1925};\ Re = \frac{d_{hb}\bar{w}}{\nu}$	
Clearance	$\Delta h_{cl} = u_2^2 0.6 \frac{\delta}{b_2} \frac{c_{2\theta}}{u_2} \times \sqrt{\frac{4\pi}{b_2 Z_{bl}} \left[\frac{r_{1t}^2 - r_{1h}^2}{(r_2 - r_{1t})(1 + \rho_2/\rho_1)}\right] \frac{c_{2\theta}}{u_2} \frac{c_{1m}}{u_2}}$		Jansen [15]
Mixing	$\Delta h_{mix} = \frac{1}{2} \frac{c_2^2}{1 + (c_{2\theta}/c_{2m})^2} \cdot \left[\frac{1 - \epsilon - B}{1 - \epsilon}\right]^2$		Johnston & Dean [17]
Disk friction	$W_{df} = f_{df} \frac{\bar{\rho} r_2^2 u_2^3}{4\dot{m}}$	$\bar{\rho} = \frac{\rho_1 + \rho_2}{2};\ f_{df} = \begin{cases} \frac{2.67}{Re_{df}^{0.5}}, & Re_{df} < 3 \cdot 10^5 \\ \frac{0.0622}{Re_{df}^{0.2}}, & Re_{df} \geq 3 \cdot 10^5 \end{cases}$	Daily & Nece [11] as quoted by Oh et al. [23]
Recirculation	$W_{rc} = 8 \cdot 10^{-5} \sinh\left(3.5\alpha_2^3\right) D_f^2 u_2^2$		Oh et al. [23]
Leakage	$W_{lk} = \frac{m_{cl} u_{cl} u_2}{2\dot{m}}$	$u_{cl} = 0.816\sqrt{2\Delta p_{cl}/\rho_2};\ \dot{m}_{cl} = \rho_2 Z_{bl} \delta L_{fl} u_{cl}$	Aungier [5]
		$\Delta p_{cl} = \frac{\dot{m}(r_2 c_{2\theta} - r_1 c_{1\theta})}{Z_{bl} \bar{r} \bar{b} L_{fl}};\ \bar{r} = \frac{r_1 + r_2}{2};\ \bar{b} = \frac{b_1 + b_2}{2}$	

Note: Equations are presented for the meridional angle convention. The nomenclature is given in SI units. B: ratio of diffuser inlet depth to impeller tip flow passage depth, b: width, c: absolute velocity, d: diameter, h: specific enthalpy, L_{ax}: axial length of impeller, $\dot{m}$: mass flow, r: radius, u: peripheral velocity, W: specific work, w: relative velocity, Z_{bl}: number of blades, α: absolute flow angle, β: relative flow angle, δ: clearance gap width, ϵ: fraction of blade-to-blade space occupied by the wake, η: dynamic viscosity, ν: kinematic viscosity, ρ: density, $(\circ)_1$: impeller inlet, $(\circ)_2$: impeller exit, $(\circ)_{bl}$: blading, $(\circ)_h$: hub, $(\circ)_m$: meridional, $(\circ)_t$: tip, $(\circ)_\theta$: tangential

4 Results

4.1 Performance Analysis

Results of the RANS-based performance calculations, previously reported in [18], and the meanline analysis for the two distinct inlet states are compared to the available experimental data reported by Wright et al. [32,34] and Fuller & Eisemann [12] in Fig. 3. The experimental performance data is interpreted to be associated with the impeller wheel, based on the instrumentation reported in [32,34] (static pressure tap at the impeller wheel exit) and the statement given in [32] that impeller wheel efficiencies (given as isentropic total-to-static efficiencies) were predicted through meanline codes and compared to experimental data. Because of high fluctuations in inlet conditions, the experimental data is provided in corrected and non-dimensionalised form. Consequently, a non-dimensionalised representation is adopted for the sake of comparison through the impeller ideal head coefficient $\psi = {(h_{2s}-h_{t1})}/{u_2^2}$ and the impeller total-to-static isentropic efficiency $\eta_s = {(h_{2s}-h_{t1})}/{(h_{t2}-h_{t1})}$, both plotted against the flow coefficient $\phi = {4\dot{V}_1}/{(\pi d_2^2 u_2)}$. Impeller exit values in CFD calculations are derived from a radial plane at $r/r_2 \approx 1.008$. This corresponds to the radial position halfway between the impeller exit radius and the vaned diffuser leading edge radius in the experimental setup.

Compared to the experimental data at the near-critical operating condition (state A), the RANS calculations show an overall flatter characteristic with higher ideal impeller head generation observed for data points at comparatively high flow coefficients and lower head generation than in the experiments towards lower flow coefficients. Noticeably, both experimental data sets share a similar qualitative shape. However, a reduced operating margin towards surge can be identified in the data set provided by Fuller & Eisemann. Since the experiments were performed with a channel diffuser, which generally restricts the flow range compared to a vaneless configuration, the flow range between experiment and CFD is not directly comparable. Range assessments are also omitted in the meanline analysis, which is presented for a setup with and without the consideration of tip clearance. Note that external (parasitic) losses are excluded in the meanline analysis for better comparability with the RANS simulations, where these are not assessed, and the experiments. For the latter, electric, magnetic and mechanical power losses estimated through correlations were excluded from the efficiency calculations, which were based on motor power measurements [32,33].

As a relatively high sensitivity towards input of the parameter ϵ was identified, which defines the fraction of blade-to-blade space occupied by the wake in the mixing loss model of Johnston & Dean [17], it is varied for three sample values for the case without tip clearance. A value of $\epsilon = 0.45$ is considered to be the most common assumption for centrifugal compressors (based on the common value of 0.2 for the wake mass fraction χ and the model formulation $\chi = \epsilon^2$, both reported in [16]) and leads to a good qualitative agreement with the RANS calculations concerning efficiency. However, in case of the head curve, a reduced value of $\epsilon = 0.35$ leads to better and satisfactory qualitative as well as quantitative agreement with the RANS calculations. The registered peak efficiency in the RANS calculations is approximately 54.4% and more than ten percentage points lower than the highest values of the highly scattered experimental data. Meanline efficiency predictions with $\epsilon = 0.45$ exceed the RANS assessment by about 5 percentage points. The consideration of tip clearance leads to decrease of 7 to 14% in head generation and a decrease in efficiency of around 4 to 6 percentage points in the meanline analysis, indicating an influential impact on compressor performance.

Comparing the datasets of the performance calculations (RANS and meanline) for the gaseous inlet state (state B) to those of the near-critical inlet state (state A), almost no difference can be observed, evidencing a high degree of machine similarity. The experimental head curve is comparable to those of the near-critical state, but shows a flatter characteristic and therefore, a better agreement between experiment and numerics is derived at this operation state. Overall, considering the high uncertainties of the test case related to the exact geometry, knowledge of the exact data assessment positions and stable process conditions in experiments, reasonable performance metrics are derived that provide a solid foundation for flow investigations.

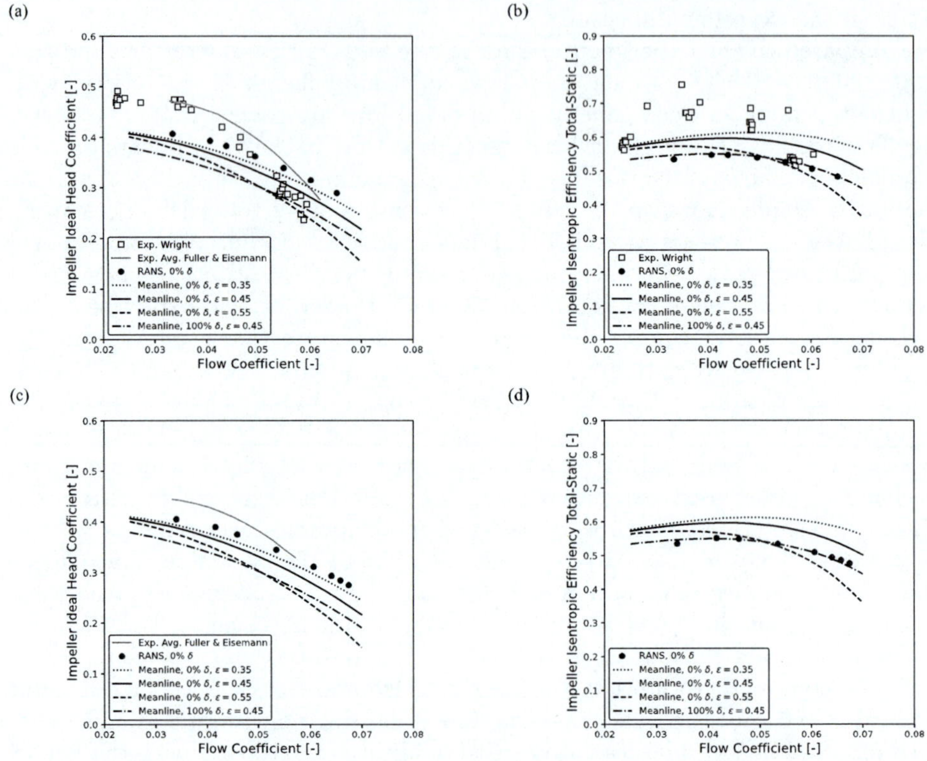

Fig. 3. Impeller performance for the total inlet condition at state A (a, b) and state B (c, d) at 50 krpm rotational speed, compared to experimental data of Wright et al. [32,34] and Fuller & Eisemann [12]. Note: δ: clearance gap width; ϵ: fraction of blade-to-blade space occupied by the wake, see also mixing loss model of Johnston & Dean [17] in Table 1.

4.2 Internal Loss Distribution Within the Meanline Analysis

The share of each considered loss mechanism in the total internal loss calculated within the meanline analysis procedure is assessed for both compressor inlet states and for different flow coefficients in Fig. 4. These calculations were performed with tip clearance and the commonly applied value of $\epsilon = 0.45$. Both charts show an almost identical loss distribution, thus supporting the indication of a high degree of machine similitude, which was illustrated in the non-dimensional performance maps. Across the entire flow spectrum, tip clearance losses are identified as a very dominant loss contributing factor, with shares of around 40%. This can be attributed to the relatively wide tip clearance gap, compared to the overall scale of the compact sCO_2 machine ($\delta/d_2 \approx 0.007$). Only for comparatively high flow coefficients, the mixing loss exceeds the clearance loss and explains the high sensitivity of the performance curves towards empirical inputs of the spatial wake fraction ϵ in these flow regions. While skin friction losses contribute with a rather constant share of around 10% of the total internal loss, blade loading losses accounting for secondary flow losses due to blade-to-blade pressure gradients are expectedly apparent for the low flow coefficient states, which overall are associated with high pressure gradients within the impeller. Considering incidence losses, it has to be taken into account that the blade angle distribution at the impeller inlet is not fully documented. Inducer blade angles were estimated on basis of the only given blade angle at the inducer tip and velocity triangle relations at design rotational speed (75 krpm), under the assumption of a constant spanwise meridional velocity and no incidence.

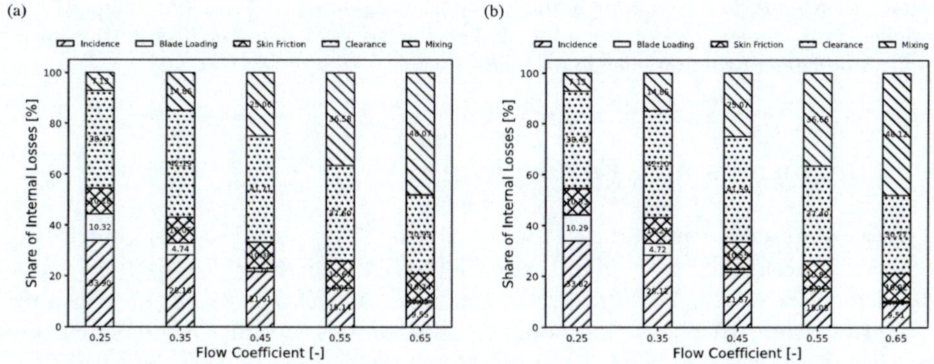

Fig. 4. Share of individual loss mechanisms in the total internal loss for different operating conditions at fixed compressor inlet total conditions at state A (a) and state B (b) at 50 krpm rotational speed.

4.3 Flow Field Analysis

Most noticeably, fluid zones with thermodynamic states within the vapour-liquid region can be identified through flow field analysis, previously reported in [18], for all operating conditions at both of the investigated compressor inlet states (exemplary shown for a single operating point in Fig. 5a). These fluid regions are observed to be located near the impeller blade leading edges and at their suction side (see Fig. 5b), where flow acceleration (see Fig. 5c) yields decreased static conditions. Volume fractions of vapour-liquid fluid regions as part of the entire computational domain were quantified for different flow coefficients and indicated a maximum share below 0.02% for state A and below 1.1% for state B, the latter being expectedly higher due to the closer location to the vapour-liquid region. However, a detailed investigation of the two-phase flow has to account for metastable states and non-equilibrium condensation of CO_2. This will be subject to extensions of the numerical framework.

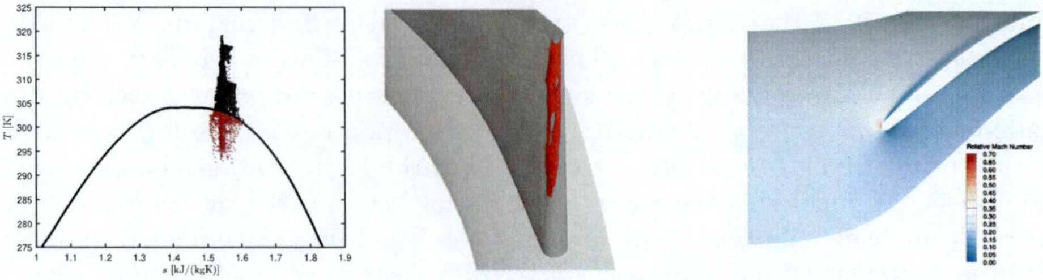

(a) (b) (c)

Fig. 5. Flow field assessments for compressor operation at state A and a flow coefficient of $\phi \approx 0.041$: (a) T,s-diagram with scattered data of the solution domain; (b) Isovolume of data points inside the vapour-liquid two-phase region near the leading edge of the impeller main blade; (c) Contour plot of the relative Mach number distribution near the leading edge of the impeller main blade.

5 Conclusions and Future Works

This work presents a numerical investigation of a supercritical CO_2 centrifugal compressor performed with an in-house CPU/GPU compressible CFD solver and complemented with a 1D single-zone meanline analysis, based on an enthalpy loss formulation. Real gas thermophysical properties within CFD calculations are accounted for by applying the Spline-Based Table Look-Up Method (SBTL), which was particularly developed for time-efficient and accurate fluid-property computations in extensive numerical simulations. Benchmarking the computational speed of the applied methodology against an ideal gas computation indicated a computational overhead of merely 33%. The candidate compressor geometry is based on the geometry of a main compressor installed in a test-loop at Sandia National Laboratories (SNL). Despite the approximation of geometry, which limits quantitative comparison against experimental data, reasonable performance metrics are obtained for compressor operation at an inlet state near the critical point as well as an inlet state in the gas phase, demonstrating the accuracy of the applied methodology. A non-dimensionalised representation of the numerically derived performance parameters shows almost no difference for both investigated compressor operation inlet conditions in comparison, which is an indication for a high degree of machine similitude. Referenced non-dimensional experimental performance data also shows comparability for the two investigated inlet states. These observations suggest that non-dimensionalised sCO_2 compressor performance testing could be numerically and practically conducted at inlet states with less pronounced gradients in thermophysical properties, than observed at operation close to the critical point. Respective findings could then be related to other inlet states through redimensionalisation. However, it has to be considered that operating states close to the vapour-liquid region might potentially lead to condensation. As a matter of fact, the resolved flow field indicated fluid regions entering the two-phase region, due to flow acceleration near the impeller blade leading edges, for all calculations.

The machine similarity is approved by comparison of the almost identical share of different loss contributing mechanisms on the total internal loss within the meanline analysis, for both inlet states and at varying flow coefficients. It is shown that losses due to tip clearance, which is relatively wide compared to the overall scale of the candidate sCO_2 machine, are a dominant loss contributing factor over the entire operating range. These are only exceeded by mixing losses for high flow coefficients. However, it is identified that the latter are very sensitive to an empirical model input factor accounting for the spatial wake fraction, which limits predictive capabilities.

Overall, the studies within this work prove that the CFD solution method, particularly tailored for the inclusion of real gas tabulation techniques, in combination with SBTL tabulation is a time-efficient and accurate methodology for complex CO_2 real gas computations in the context of turbomachinery. These can be complemented with the presented meanline analysis method considering certain restrictions owed to empirical user input. Future work will focus on integration of metastable states and the assessment of non-equilibrium condensation. Furthermore, the geometry will be extended to account for the channel diffuser in order to improve the prediction capabilities regarding the compressor operating range.

References

1. Ahn, Y., et al.: Review of supercritical CO_2 power cycle technology and current status of research and development. Nucl. Eng. Technol. **47**(6), 647–661 (2015)
2. Ameli, A., Turunen-Saaresti, T., Backman, J.: Numerical investigation of the flow behavior inside a supercritical CO_2 centrifugal compressor. J. Eng. Gas Turbines Power **140**(12), 122604-1–122604-7 (2018)
3. Ameli, A., et al.: Compressor design method in the supercritical CO_2 applications. In: Supercritical CO_2 Power Cycles Symposium, Pittsburgh, Pennsylvania, USA (2018)
4. Ameli, A., et al.: Effects of real gas model accuracy and operating conditions on supercritical CO_2 compressor and flow field. J. Eng. Gas Turbines Power **140**(6), 062603-1–062603-8 (2018)
5. Aungier, R.H.: Mean streamline aerodynamic performance analysis of centrifugal compressors. J. Turbomach. **117**(3), 360–366 (1995)
6. Bell, I., et al.: Pure and pseudo-pure fluid thermophysical property evaluation and the open-source thermophysical property library CoolProp. Ind. Eng. Chem. Res. **53**(6), 2498–2508 (2014)
7. Concepts NREC LLC. AxCent64 (8.7.13.0) (2018)
8. Concepts NREC LLC. COMPAL (8.7.13.0) (2018)

9. Conrad, O., Raif, K., Wessels, M.: The calculation of performance maps for centrifugal compressors with vane-island diffusers. Proceedings of the Twenty-Fifth Annual International Gas Turbine Conference and Exhibit and Twenty-second Annual Fluids Engineering Conference, New Orleans, Louisiana, USA (1979)
10. Coppage, J.E., et al.: Study of supersonic radial compressors for refrigeration and pressurization systems. WADC TECHNICAL REPORT 55-257, Airesearch Manufacturing Company (1956)
11. Daily, J.W., Nece, R.E.: Chamber dimension effects on induced flow and frictional resistance of enclosed rotating disks. J. Basic Eng. **82**(1), 217–230 (1960)
12. Fuller, R.L., Eisemann, K.: Centrifugal compressor off-design performance for super-critical CO_2. In: Supercritical CO_2 Power Cycles Symposium, Boulder, Colorado, USA (2011)
13. Hathaway, M.D., et al.: Laser anemometer measurements of the three-dimensional rotor flow field in the NASA low-speed centrifugal compressor. Army Research Laboratory Technical Report ARL-TR-333, NASA (1995)
14. Iseni, S., Post, P., di Mare, F.: Numerical analysis of the influence of air humidity on a transonic compressor stage. In: International Gas Turbine Congress 2019. Paper No. ThAM16.1, Tokyo (2019)
15. Jansen, W.: A method for calculating the flow in a centrifugal impeller when entropy gradients are present. In: Royal Society Conference on Internal Aerodynamics (Turbomachinery), Cambridge, UK (1967)
16. Japikse, D.: Centrifugal Compressor Design and Performance. Concepts ETI (1996)
17. Johnston, J.P., Dean Jr., R.C.: Losses in vaneless diffusers of centrifugal compressors and pumps: analysis, experiment, and design. J. Eng. Power **88**(1), 49–60 (1966)
18. Karaefe, R.E., et al.: Numerical investigation of a centrifugal compressor for supercritical CO_2 cycles. In: Proceedings of the ASME Turbo Expo 2020. ASME Paper No. GT2020-15194 (2020)
19. Kunick, M.: Fast calculation of thermophysical properties in extensive process simulations with the Spline-Based Table Look-Up Method (SBTL). Fortschritt-Berichte VDI. Nr. 618, Reihe 6, Energietechnik (2018)
20. Kunick, M., et al.: CFD analysis of steam turbines with the IAPWS standard on the Spline-Based Table Look-Up Method (SBTL) for the fast calculation of real fluid properties. In: Proceedings of the ASME Turbo Expo 2015. ASME Paper No. GT2015-43984 (2015)
21. Lee, J., et al.: Design methodology of supercritical CO_2 Brayton cycle turbomachineries. In: Proceedings of the ASME Turbo Expo 2012. ASME Paper No. GT2012-68933 (2012)
22. NUMECA International. AutoGrid5 (13.1) (2018)
23. Oh, H.W., Yoon, E.S., Chung, M.K.: An optimum set of loss models for performance prediction of centrifugal compressors. Proc. Inst. Mech. Eng. Part A J. Power Energy **211**(4), 331–338 (1997)
24. Post, P., di Mare, F.: Highly efficient Euler-Euler approach for condensing steam flows in turbomachines. In: GPPS Montreal 18, Montreal, Canada (2018)
25. Post, P., Sembritzky, M., di Mare.: Towards scale resolving computations of condensing wet steam flows. In: Proceedings of the ASME Turbo Expo 2019. ASME Paper No. GT2019-91269 (2019)
26. Post, P., Winhart, B., di Mare, F.: Large eddy simulation of a condensing wet steam turbine cascade. In: Proceedings of the ASME Turbo Expo 2020. ASME Paper No. GT2020-16064 (2020)

27. Romei, A., et al.: The role of turbomachinery performance in the optimization of supercritical carbon dioxide power systems. J. Turbomach. **142**(7), 071001-1–071001-11 (2020)
28. Spalart, P.R., Allmaras, S.R.: A one-equation turbulence model for aerodynamic flows. In: 30th Aerospace Sciences Meeting and Exhibit, Reno, Nevada, USA (1992)
29. Span, R., Wagner, W.: A new equation of state for carbon dioxide covering the fluid region from the triple-point temperature to 1100 K at pressures up to 800 MPa. J. Phys. Chem. Ref. Data **25**(6), 1509–1596 (1996)
30. van Leer, B.: Towards the ultimate conservative difference scheme. V. A second-order sequel to Godunov's method. J. Comput. Phys. **32**(1), 101–136 (1979)
31. Wiesner, F.J.: A review of slip factors for centrifugal impellers. J. Eng. Gas Turbines Power **89**(4), 558–566 (1967)
32. Wright, S.A., et al.: Supercritical CO_2 compression loop operation and test results. In: Supercritical CO_2 Power Cycles Symposium, Troy, New York, USA (2009)
33. Wright, S.A., et al.: Supercritical CO_2 compression loop operation and test results. Presentation. Supercritical CO_2 Power Cycles Symposium, Troy, New York, USA (2009)
34. Wright, S.A., et al.: Operation and analysis of a supercritical CO_2 Brayton cycle. SANDIA REPORT SAND2010-0171, Sandia National Laboratories (2010)

Rigs and Components

Design of a Closed-Loop Supercritical Carbon Dioxide Wind Tunnel: Numerical Modelling of Non-equilibrium Condensation in a Converging-Diverging Nozzle

Giuseppe Petruccelli(✉), Antti Uusitalo, Aki Grönman, Teemu Turunen-Saaresti, and Marta Zocca

Laboratory of Fluid Dynamics, Lappeenranta University of Technology, Lappeenranta 53850, Finland
{Giuseppe.Petruccelli,Antti.Uusitalo,Aki.Gronman, Teemu.Turunen-Saaresti,Marta.Zocca}@lut.fi

Abstract. Closed-loop supercritical CO_2 (sCO_2) cycles are one of most promising power conversion options for the foreseeable future. Experimental studies are necessary for a successful development and deployment of these technologies. This paper focuses on the design of a sCO_2 experimental loop with a supersonic nozzle test section. The expansion process has operating conditions similar to those found at the inlet of supercritical compressors with non-equilibrium condensation. Among the possible cycle configurations, the transcritical refrigeration cycle is preferred for its wide range of applications and the operability in the thermodynamic region of interest close to the critical point. The design of the converging-diverging nozzle is also discussed. The saturation curve is crossed in the diverging section of the nozzle, therefore a non-equilibrium condensation model is presented and validated for an accurate design of the test section. In addition, the validated model is employed to present first results from the designed nozzle.

Keywords: Supercritical carbon dioxide · Transcritical refrigeration cycle · Supersonic nozzle · Condensation

1 Introduction

Supercritical CO_2 (sCO_2) cycles offer a competitive opportunity in several industrial and power generation applications. The high compactness of the components and the high efficiencies make these systems very attractive for applications such as fossil, nuclear, concentrating solar power (CSP) plants and waste heat recovery (Angelino 1968; Dostal et al. 2006; Allam et al. 2017; Uusitalo, et al. 2018). Furthermore, CO_2 is a safe and economic fluid, which makes it very popular in applications such as industrial refrigeration and heat pumps

M. Pini et al. (Eds.): NICFD 2020, ERCOFTAC Series 28, pp. 103–113, 2021.
https://doi.org/10.1007/978-3-030-69306-0_11

(Groll et al. 2007; Austin et al. 2011). Its low critical temperature allows this type of cycles to be operated transcritically with pressures up to 130 bar, solving the problems that subcritical systems encounter when operating with heat rejection temperatures close to the critical point. In sCO_2 Brayton cycle, maximum efficiencies are reached when the compressor operates near the critical point (Wright 2010). However the flow acceleration that takes place on the suction side of the blade could cause the onset of condensation, with effects on the machine performance (Baltadjiev 2015; Ameli 2018). Lettieri et al. (2018) experimentally assessed the onset of condensation for a rapid expansion in proximity of the saturation curve and defined a criterion to determine a limit for condensation-free inlet conditions of compressors. Bakhtar et al. (2005) reviewed the classical theory of the homogenous nucleation of water droplets, focusing on the most relevant modifications proposed over the years. The supercritical fluids show large deviations from the ideal-gas thermodynamic behavior, particularly approaching the critical point and the liquid-vapor saturation curve. This study presents a new test loop, which allows to study both supersonic sCO_2 expansions and to assess heat transfer properties crossing the pseudocritical line. The present work describes the preliminary design of the test loop according to the technical requirements and limitations. The layout of the thermodynamic cycle is presented in Sect. 2. In Sect. 3, the design of the components of the supersonic test section is discussed and the numerical model for real-gas non-equilibrium condensation applied to the design of the test section nozzle is presented and validated against the experimental data from Lettieri et al. (2018).

2 Thermodynamic Cycle Design

Two different test-rig configurations were considered, namely the closed-loop Brayton cycle and the transcritical refrigeration cycle. Both cycle layouts allow to operate the test section in the region of interest close to the critical point, where the real-gas behavior is relevant. The overall assessment of the cycle must take into account the constraints expressed in terms of plant size and power required. The overall power requirement is set to not exceed ~500 kW and, due to the limited space available in the laboratory, each component has to be designed minimizing the dimensions. However, the cycle must be able to provide the necessary flow rate for reaching choking conditions at the throat of the supersonic test section, which is set equal to 1.60 kg/s in design conditions. The throat area is fixed to 53.58 mm^2, in order to guarantee a throat height of 13.4 mm, which is estimated to be wide enough to allow an optical visualization of the flow. The maximum achievable throat dimensions have been calculated within the constraints of the aforementioned power limitation and the minimum admissible width of the nozzle, which makes the presence of the boundary layers negligible, i.e. the fluid motion can be considered a two-dimensional flow. In the following section, the transcritical refrigeration cycle configuration is described and analyzed considering the previously mentioned constraints. This cycle configuration is preferred because of the market availability of the components.

Most notably, the chosen compressor is a semi-hermetic reciprocating compressor commonly used in refrigeration and air conditioning units.

2.1 Transcritical Refrigeration Cycle Configuration

The transcritical refrigeration cycle configuration is represented in Fig. 1 together with the corresponding T-s diagram. The flow is compressed from superheated to supercritical conditions ($1 \rightarrow 2$). The CO_2 is then cooled down ($2 \rightarrow 3$) and the heat released is used to complete the evaporation of the flow directed to the compressor ($8 \rightarrow 1$). The thermodynamic conditions at the inlet of the test section (point 5) are controlled by an expansion valve ($3 \rightarrow 4$) and by the heater ($4 \rightarrow 5$). The test section can operate within a range of inlet temperatures and pressures of $40 \div 65$ °C and $80 \div 90$ bar respectively, with the aim of covering the region of interest depicted in Fig. 1. When the test section is open, the shock occurring in the diverging section induces a total pressure loss ($5 \rightarrow 6ts$). This loss must be carefully evaluated to ensure a proper test loop operation. The flow leaving the test section is then cooled down ($6ts \rightarrow 7ts$) before undergoing a second expansion that sets the evaporation pressure ($7ts \rightarrow 8$). Table 1 reports the main operating parameters for this configuration, where W_C is the power required from the compressor, Q_H is the heat supplied in the heater and Q_C is the heat released in the cooler.

Table 1. Main operating parameters for the transcritical refrigeration cycle configuration

Parameter	P_1	T_1	P_2	T_2	P_5	T_5	T_7	T_8	W_C	Q_H	Q_C
Value	64.3	35	130	94.8	90	65	31.9	25	64	203	261
Unit	Bar	°C	Bar	°C	Bar	°C	°C	°C	kW	kW	kW

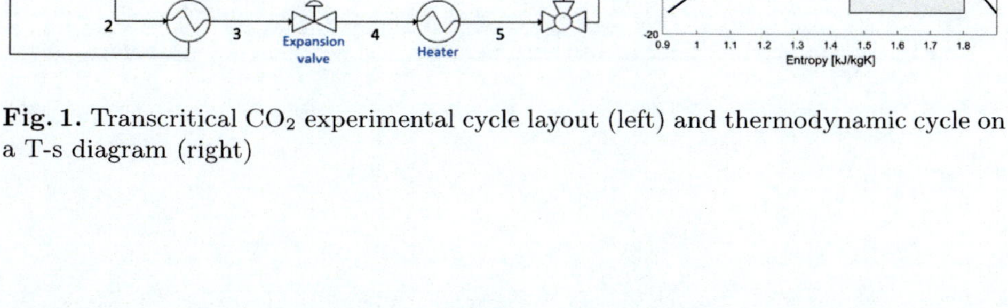

Fig. 1. Transcritical CO_2 experimental cycle layout (left) and thermodynamic cycle on a T-s diagram (right)

3 Test Section Design

As schematized in Fig. 2, the supersonic test section consists of three main components: the contraction zone, the converging-diverging nozzle and the diffuser. In the contraction zone, the flow is accelerated up to the required Mach number at the inlet of the converging-diverging nozzle and completes a transition from the circular cross-section of the upstream plenum to the rectangular cross-section of the nozzle. A planar configuration is indeed adopted in order to ensure a two-dimensional flow field within the nozzle. The flow is then accelerated to supersonic conditions in the converging-diverging section. In the diverging part of the nozzle, the thermodynamic properties cross the saturation curve, causing the nucleation of small droplets and their consequent growth. A supersonic diffuser is installed downstream of the nozzle to partially recover the kinetic energy available in this point. Different configurations for this component were considered in the preliminary design phase. Assuming a shock-free test section, the recovery of the kinetic head can be achieved by a second converging-diverging section, which would slow the flow down to subsonic conditions. However, this arrangement would introduce significant issues, particularly in the startup of the loop and in terms of operational flexibility. The alternative solution adopted in this work is to avoid a supersonic diffuser and design the divergent section of the primary nozzle by including the presence of a normal shock wave, as shown in Fig. 2. The shock wave produces a near-instantaneous deceleration of the flow to subsonic speed, leading to an increase in pressure, temperature and entropy. Finally, a diverging duct is placed at the exit of the test section to recover the rest of the kinetic head and to perform the transition from a rectangular to a circular cross-section. In the following Sect. 3.1 and 3.2, the performance of the contraction zone and the supersonic nozzle are investigated through real-gas CFD simulations. The simulations are performed with Ansys Fluent (contraction zone) and Ansys CFX (nozzle). In both solvers, thermodynamic properties are computed using lookup tables extracted from NIST REFPOP database (Lemmon et al. 2018) and are based on the Span-Wagner equation of state (Span et al. 1996). In Sect. 3.2, the condensation model is presented, validated and applied to the CFD simulation of the designed nozzle.

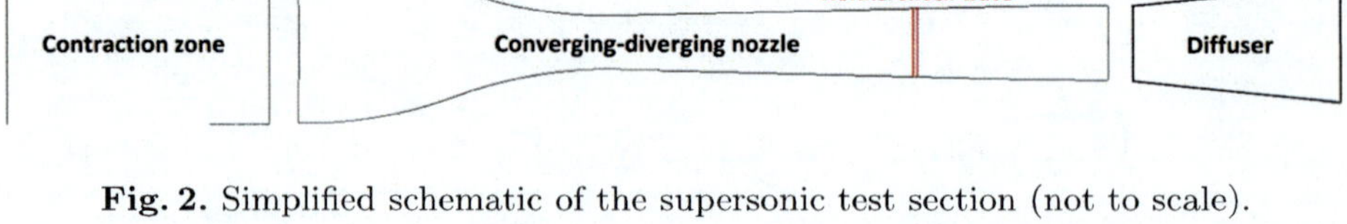

Fig. 2. Simplified schematic of the supersonic test section (not to scale).

3.1 Contraction Zone

The first component of the test section is the contraction zone, which performs two major tasks: completing the transition from the circular to the rectangular cross-section and accelerating the flow up to the designed Mach number ($M \sim 0.2$) at the inlet of the nozzle converging section. The inlet diameter of the contraction zone is equal to 50 mm and, under the assumption of isentropic process, the dimensions calculated at the outlet section are 4×40 mm. The axial length of the contraction zone, L_c, is set to 100 mm, which corresponds to twice the inlet diameter. This length is considered large enough to ensure a smooth geometry transition and to prevent flow separation along the contraction wall. The design of this component is performed using a method similar to the one developed by Passmann et al. (2016). The method foresees a transition from a circular to a rectangular cross-section by applying the superellipse, which is defined as:

$$\left|\frac{y}{a}\right|^n + \left|\frac{z}{b}\right|^n = 1 \tag{1}$$

where y and z are the cross-section coordinates and a and b are the minor and major axis of the superellipse. The shape of the ellipse is determined by the exponent n. When $n = 2$, the superellipse has the shape of a circle, whereas when $n \rightarrow \infty$ it tends to be a rectangle. Thus, the geometry transition is driven by the variation of the exponent n as a function of x/L_c, where x is the coordinate in the streamwise direction that goes from 0 to L_c. In this work, the same exponential function defined by Passmann et al. (2016) is used:

$$n\left(\frac{x}{L_c}\right) = 2 + 3.845 \cdot 10^{-3} \cdot \exp\left(10.148 \cdot \frac{x}{L_c}\right) \tag{2}$$

This function allows to obtain $n = 2$ at the inlet ($x/L_c = 0$) and $n = 100$ at the contraction exit ($x/L_c = 1$). To define the streamwise contraction, Bell et al. (1988) proposed some polynomial functions for the axial variation of the half axes $a\,(x)$ and $b\,(x)$. The fifth-order polynomial function is used in this work, which takes the form:

$$a\left(\frac{x}{L_c}\right) = a_{in} + (a_{out} - a_{in})\left|6\left(\frac{x}{L_c}\right)^5 - 15\left(\frac{x}{L_c}\right)^4 + 10\left(\frac{x}{L_c}\right)^3\right| \tag{3}$$

$$b\left(\frac{x}{L_c}\right) = b_{in} + (b_{out} - b_{in})\left|6\left(\frac{x}{L_c}\right)^5 - 15\left(\frac{x}{L_c}\right)^4 + 10\left(\frac{x}{L_c}\right)^3\right| \tag{4}$$

The first derivative of this polynomial function is set to zero for $z = 0$ and $z = L_c$ to ensure a continuous variation in the cross-section area in these locations. A 3D inviscid simulation is run to verify that no flow separation occurs in the near-wall region. Only a quarter of the geometry is considered due to symmetries and the domain at the contraction outlet is extended by 10% of the total length with a constant cross-section area duct. Figure 3 reports the Mach number calculated at the centerline and along the intersections of the wall with the symmetry planes and the diagonal plane. Some small fluctuations are observed in the exit duct, which are due to irregularities of the unstructured mesh used to better approximate this geometry. The performance of the contraction zone can be evaluated through the total pressure loss coefficient defined as:

$$Y = \frac{P_{0,in} - P_{0,out}}{P_{0,out} - P_{out}} \tag{5}$$

The total pressure loss coefficient calculated for the inviscid simulation is $Y = 0.0004$, instead by running a 3D-viscous CFD simulation this increases up to $Y = 0.0348$, which is consistent with the higher total pressure drop due to the viscous losses. These results confirm that the contraction zone is properly designed and well performing, thus ensuring a uniform flow field at the inlet of the nozzle.

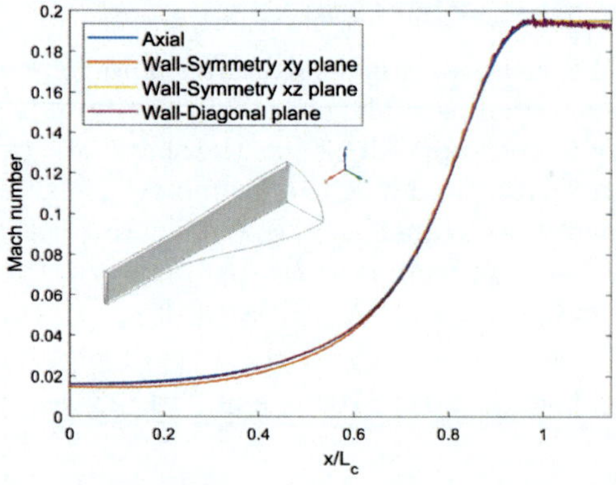

Fig. 3. Axial and wall velocity distribution from 3D-inviscid simulation

3.2 Converging-Diverging Nozzle and Validation of the Condensation Model

The converging-diverging nozzle performs the CO_2 expansion to supersonic conditions. The flow enters the converging section with an inlet Mach number of approximately 0.2 and reaches choking conditions at the throat ($M = 1$). In the diverging section, the flow expands to supersonic speeds and crosses the vapor-liquid saturation curve, triggering the nucleation of small droplets. The nozzle is designed to operate with a normal shock wave in the diverging section. The pre-shock Mach number, calculated under the assumption of equilibrium condensation, is 1.35. The total pressure losses across the shock are calculated in a pre-design phase by numerically solving the normal shock relations, complemented with the Span-Wagner equation of state. In Fig. 4, the total pressure drop calculated from the shock relations for different values of the post-shock pressure is compared with the losses computed from the two-phase CFD simulation at the design point, i.e. for $P_{0,in}$ /P_{out} $\cong$ 1.31. In design conditions, the total pressure losses reach a value of 7 bar. This value, which corresponds to 7.8% of the inlet total pressure, falls well within the limits of cycle operability.

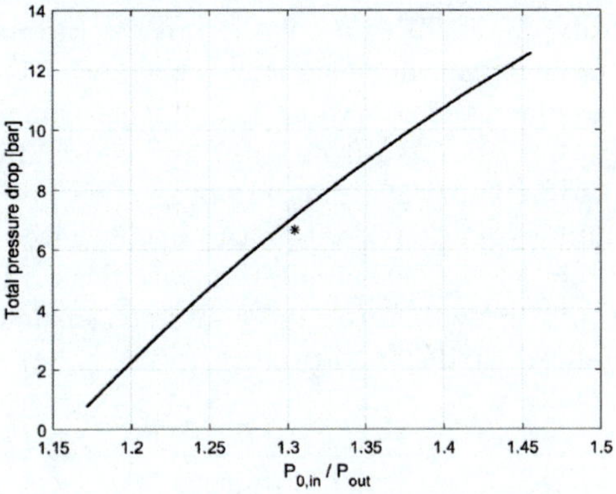

Fig. 4. Total pressure drop calculated for different back pressures. 1D code (line) and 2-phase CFD simulation (*).

The implementation of an optical measurement system is currently being investigated. Since such analyses require optical access to the test section, efforts were made to design a nozzle with the largest possible cross-sectional area (the throat height was previously estimated to be 13.4 mm). In order to perform a proper design of this component, a non-equilibrium condensation model is presented and validated with the experimental data from Lettieri et al. (2018). The homogeneous nucleation rate is computed using the classical nucleation theory of Becker and Döring with the Kantrowitz's non-isothermal correction (Kantrowitz 1950):

$$J = \frac{1}{1+\theta}\frac{\rho_g^2}{\rho_l}\sqrt{\frac{2\sigma}{\pi M^3}} \cdot \exp\left(-\frac{4\pi r^{*2}\sigma}{3k_B T_g}\right) \tag{6}$$

where ρ is the density, T is the temperature, σ is the surface tension, M is the molecular mass, k_B is the Boltzmann constant, r^* is the critical radius, θ is the non-isothermal correction term and the subscripts g and l stands for the gas and liquid phase respectively. The surface tension is defined by a polynomial regression derived from the NIST REFPROP values. The critical radius is computed taking into account the compressibility factor Z, which is less than 1 near the saturation curve. For a given saturation pressure, a linear dependence of the compressibility factor on pressure is assumed. This approximation can be considered valid in proximity of the saturation curve and therefore the critical radius can be expressed as:

$$r^* = \frac{2\sigma}{\rho_l R T_g \int_{P_s}^{P} \frac{Z}{P} dP} = \frac{2\sigma}{\rho_l R T_g \left[\ln S \ + A\left(P_s\right) \cdot P_s \left(S-1\right)\right]} \tag{7}$$

where S is the supersaturation ratio, defined as the ratio between the pressure and the saturation pressure of the gas phase (P/P_s), and the coefficient A is calculated through a second degree polynomial function of the saturation pressure. CFX automatically clips the value of the critical radius to 1 mm to avoid numerical instability issues near the saturation curve ($S \sim 1$). To overcome this problem, a "dummy" surface tension value is defined through a linear function of the correct critical radius obtained from Eq. 7 and is used by the solver only to calculate the critical radius value avoiding clipping. The interfacial mass transfer between the continuous and dispersed phases is computed in CFX through a droplet growth rate model. The CFD simulations are performed employing both the CFX standard droplet growth rate model and a user-defined one, hereafter referred to as the enhanced model (Gyarmathy 1963):

$$\frac{dr}{dt} = \frac{k_g}{r \rho_l \left[\frac{1}{1+4Kn} + 3.78 \left(1-\nu\right) \frac{Kn}{Pr_g} \right]} \left(\frac{T_l - T_g}{L} \right) \tag{8}$$

where k is the thermal conductivity, Kn is the Knudsen number, L is the specific enthalpy of evaporation, Pr is the Prandtl number and ν is the Young modification parameter (Young 1982). The droplet temperature, T_l, is calculated in the solver using a simple algebraic equation:

$$T_l = T_s\left(P\right) - \left[T_s\left(P\right) - T_g\right] \frac{r^*}{r} \tag{9}$$

The solution of the energy equation for the dispersed phase is not required in this simplified approach, thus reducing the complexity of the simulation and allowing for a faster convergence.

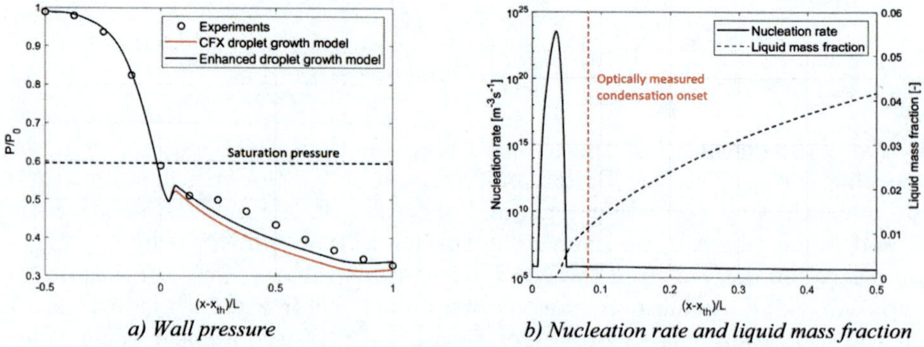

Fig. 5. Wall pressure distribution (left), nucleation rate and liquid mass fraction calculated with the enhanced droplet growth rate model (right). Nozzle from Lettieri et al. (2018).

The simulation results reported in Fig. 5 show good agreement with the experiments in terms of wall pressure distribution, particularly when the enhanced droplet growth model is used, even though the calculated onset of condensation is shifted upstream by about 3 mm with respect to the optically measured one. In the light of the results obtained, the condensation model with the enhanced droplet growth model is employed in the CFD simulation of the designed nozzle. The total pressure and total temperature at the inlet boundary are set to 90 bar and 65 °C respectively, coherently with the thermodynamic cycle previously discussed (Fig. 1). The expansion up to the shock wave and the subsequent condensation are represented in Fig. 6a in a T-s diagram. The compressibility factor map, superposed on the T-s diagram, emphasizes the real-gas behavior of the fluid, which was considered in the definition of the condensation model. The fluid crosses the saturation curve with an expansion rate of 2.1 $\cdot\ 10^3\ s^{-1}$ and once it reaches the Wilson point, the pressure and temperature rapidly return to saturated conditions. The wall pressure distribution in the nozzle is reported in Fig. 6b for two cases. The black and the red lines represent the cases with and without the condensation model, respectively. Figure 6c shows the nucleation rate and the liquid mass fraction calculated on the axial centerline of the nozzle. The CO_2 is in a superheated state downstream of the shock, causing a rapid decay of liquid droplets. At this point, the flow is directed to the diffuser to recover the remaining kinetic head.

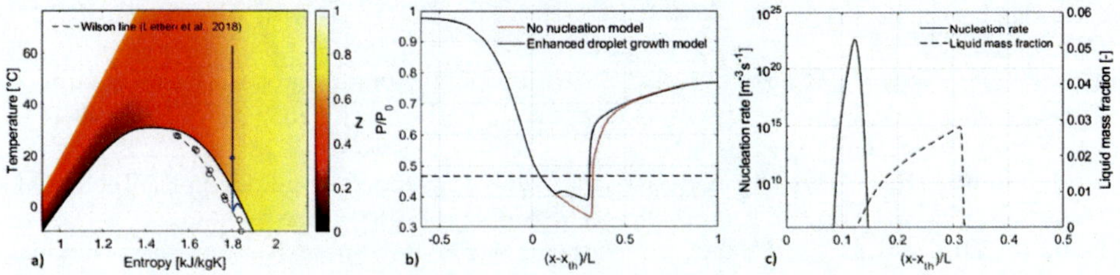

Fig. 6. Designed nozzle: (a) Expansion and condensation in the T-s diagram, superposed to the compressibility factor map. The symbol * indicates the throat section (M = 1). (b) Wall pressure distribution, (c) nucleation rate and liquid mass fraction.

4 Conclusions and Future Works

The design of a closed-loop supercritical carbon dioxide wind tunnel is carried out. The cycle layout is discussed and designed according to the technical requirements of the test section and the constraints on power and dimensions. The transcritical refrigeration cycle layout is preferred for its wide use in various industrial applications and its feasibility in terms of plant size and economic constraints. The advantage of this experimental loop is the capacity of working continuously with operating parameters located in the supercritical region and particularly close to the critical point, where there is still little literature available, both in terms of fundamentals and experimental data. The numerical modelling of the non-equilibrium condensation is performed through real-gas CFD simulations carried out in Ansys CFX. The validated model is applied to properly design the test section that will be part of the experimental setup. The construction and assembly of the experimental loop is foreseen by the end of 2020.

References

Allam, R., Scott, M., Forrest, B., Fetvedt, J., Lu, X., Freed, D., Brown, G.W., Sasaki, T., Itoh, M., Manning, J.: Demonstration of the Allam Cycle: an update on the development status of a high efficiency supercritical carbon dioxide power process employing full carbon capture. Energy Procedia **114**, 5948–5966 (2017). https://doi.org/10.1016/j.egypro.2017.03.1731

Ameli, A., Turunen-Saaresti, T., Backman, J.: Numerical investigation of the flow behavior inside a supercritical CO2 centrifugal compressor. ASME J. Eng. Gas Turbines Power **140**(12), 122604 (2018). https://doi.org/10.1115/GT2016-57481

Angelino, G.: Carbon dioxide condensation cycles for power production. ASME J. Eng. Power **90**(3), 287–295 (1968). https://doi.org/10.1115/1.3609190

Austin, B.T., Sumathy, K.: Transcritical carbon dioxide heat pump systems: a review. Renew. Sustain. Energy Rev. **15**, 4013–4029 (2011). https://doi.org/10.1016/j.rser.2011.07.021

Bakhtar, F., Young, J.B., White, A.J., Simpson, D.A.: Classical nucleation theory and its application to condensing steam flow calculations. Proc. Inst. Mech. Eng. Part C J. Mech. Eng. Sci. **219**(12), 1315–1333 (2005). https://doi.org/10.1243/095440605X8379

Baltadjiev, N., Lettieri, C., Spakovszky, Z.: An investigation of real gas effects in supercritical CO2 centrifugal compressors. ASME J. Turbomach. **137**(9), 091003 (2015). https://doi.org/10.1115/1.4029616

Bell, J., Mehta, R.: Contraction design for small low-speed wind tunnels. Technical report, NASA Contractor Report No. NASA–CR–177488 (1988)

Dostal, V., Driscoll, M.J., Hejzlar, P.: High-performance supercritical carbon dioxide cycle for next-generation nuclear reactors. Nucl. Technol. **154**(3), 265–282 (2006). https://doi.org/10.13182/NT154-265

Groll, E.A., Kim, J.H.: Review of recent advances toward transcritical CO2 cycle technology. HVAC&R Res. **13**(3), 499–520 (2007). https://doi.org/10.1080/10789669.2007.10390968

Gyarmathy, G.: Zur Wachstumsgeschwindigkeit kleiner Flüssigkeitstropfen in einer übersättigten Atmosphäre. Zeitschrift Für Angewandte Mathematik Und Physik ZAMP **14**(3), 280–293 (1963). https://doi.org/10.1007/BF01601066

Kantrowitz, A.: Nucleation in very rapid vapor expansion. J. Chem. Phys. **19**(9), 1097–1100 (1950). https://doi.org/10.1063/1.1748482

Lemmon, E.W., Bell, I.H., Huber, M.L., McLinden, M.O.: NIST Standard Reference Database 23: Reference Fluid Thermodynamic and Transport Properties-REFPROP, Version 10.0. National Institute of Standards and Technology, Standard Reference Data Program, Gaithersburg (2018). https://doi.org/10.18434/T4JS3C

Lettieri, C., Paxson, D., Spakovszky, Z., Bryanston-Cross, P.: Characterization of nonequilibrium condensation of supercritical carbon dioxide in a de laval nozzle. ASME J. Eng. Gas Turbines Power **140**(4), 041701 (2018). https://doi.org/10.1115/1.4038082

Passmann, M., Reinker, F., Hasselmann, K., auf der Wiesche, S., Joos, F.: Development and design of a two-stage contraction zone and test section of an organic rankine cycle wind tunnel. In: ASME Turbo Expo, Seoul, South Korea, GT2016-56580, 13–17 June 2016 (2016). https://doi.org/10.1115/GT2016-56580

Span, R., Wagner, W.: A new equation of state for carbon dioxide covering the fluid region from the triple-point temperature to 1100 K at pressure up to 800 MPa. J. Phys. Chem. Ref. Data **25**, 1509–1596 (1996). https://doi.org/10.1063/1.555991

Uusitalo, A., Ameli, A., Turunen-Saaresti, T.: Thermodynamic and turbomachinery design analysis of supercritical Brayton cycles for exhaust gas heat recovery. Energy **167**, 60–79 (2018). https://doi.org/10.1016/j.energy.2018.10.181

Wright, S.A., Radel, R.F., Vernon, M.E., Rochau, G.E., Pickard, P.S.: Operation and Analysis of a Supercritical CO2 Brayton Cycle. Sandia National Laboratories, Albuquerque, NM, Sandia Report No. 2010-0171 (2010). https://doi.org/10.2172/984129

Young, J.B.: The spontaneous condensation of steam in supersonic nozzles. Physico-Chemical Hydrodyn. (PCH) **3**, 57–82 (1982). ISSN 0191-9059

Measurement System of Small-Scale High Expansion Ratio ORC Turbine

Antti Uusitalo(✉), Marta Zocca, and Teemu Turunen-Saaresti

LUT University, Lappeenranta, Finland
{antti.uusitalo,marta.zocca,teemu.turunen-saaresti}@lut.fi

Abstract. Small-scale and high temperature ORCs have received increasing interest in recent times. This type of power systems can have high expansion ratio turbines with gas dynamic effects differing significantly from the behaviour of ideal gases, as the turbines are often operated in the non-ideal dense-gas region. One of the most crucial steps leading to future improvement of turbine efficiency and to the development of design guidelines and loss analysis methods is to validate the applied design methods and loss correlations with experimental data. However, accurately measuring the performance of ORC turbines can be challenging. Indeed, despite the high pressure ratios, the temperature drop across the turbines remains rather low, and the physical size of turbomachinery tends to be small. In this paper, updates to the measurement system of the 10 kW LUT micro-ORC test-rig operating with siloxane MDM are presented. Further measurement points are placed at turbine inlet and outlet, and inside the turbine stator, with the aim of obtaining accurate performance maps and turbine loss analysis from future experimental runs.

Keywords: ORC turbine · Turbine measurements · Dense gas flows

1 Introduction

Small-scale (about 10 kW to 50 kW) ORCs have been intensively studied in recent times. Especially in high-temperature small-scale ORCs, fluids with high molecular weight and with relatively high critical temperature are adopted. This type of fluids are characterized by gas dynamic effects that differ significantly from those of ideal gases, especially if the turbine of the system is designed to operate in the dense-gas region. The use of high molecular weight and critical temperature fluids in ORCs also often results in high expansion ratio over the turbine, whereas simultaneously the temperature drop remains moderate. In addition, as the speed of sound of these fluids is rather low, the turbine stator has to be designed for highly supersonic flow conditions [1]. There are not many experimental studies on small-scale high temperature ORC turbines in

M. Pini et al. (Eds.): NICFD 2020, ERCOFTAC Series 28, pp. 114–122, 2021.
https://doi.org/10.1007/978-3-030-69306-0_12

NOMENCLATURE

Roman symbols

p	Pressure
q_m	Flow rate
T	Temperature

Greek symbols

Γ	Fundamental derivative of gas dynamics
Z	Compressibility factor

Sub- and super-indices

crit	Critical
des	Design point
out	Outlet

the present literature, see e.g.. Ref. [2,7]. Thus, it is crucial to carry out more experimental research on small ORC turbines. Due to the gas dynamic effects of organic fluids, it is still unclear how such turbines should be designed in respect to the more conventional turbomachinery design guidelines. In addition, it is still unclear what is the contribution of different individual loss sources on the overall losses of ORC turbines.

In this paper, the updated turbine measurement system of the LUT micro ORC test setup is presented. The system has a 10 kW scale highly supersonic radial turbine, and uses siloxane MDM as a working fluid. The turbine inlet pressure of 7.9 bar, temperature of 265 °C, fluid mass flow rate of 0.2 kg/s and turbine outlet pressure of about 0.07 bar were defined as the turbine design operating conditions. More details on the operating points of the setup and the working fluid selection are described in Ref. [4]. Several test were run with the facility, and the system turbogenerator was successfully operated with the design operating conditions with rotational speeds of up to 30 000 rpm, see e.g. Ref. [5]. However, in the experiments carried out so far, it was observed that accurately defining the turbine performance during the experiments is sensitive especially on the accuracy of turbine outlet state measurements, namely the turbine outlet temperature and pressure measurement, and working fluid mass flow rate measurement. The specific thermodynamic characteristics of siloxane MDM, including a low or moderate temperature change even with high turbine pressure ratios, highlights the need for accurately measuring the thermodynamic state at the turbine inlet and outlet. It was concluded that the original measurement system was not accurate enough to accurately estimate the turbine performance in the tests. Thus, an improved measurement system is now adopted, which includes improved turbine outlet state measurement and mass flow rate measurements. In addition, pressure measurement points are added to turbine stator nozzles in order to estimate the operation and performance of the stator more accurately, as well as to evaluate how the expansion is divided between the stator and rotor during the experiments. The overall objective of the updates to the measurement

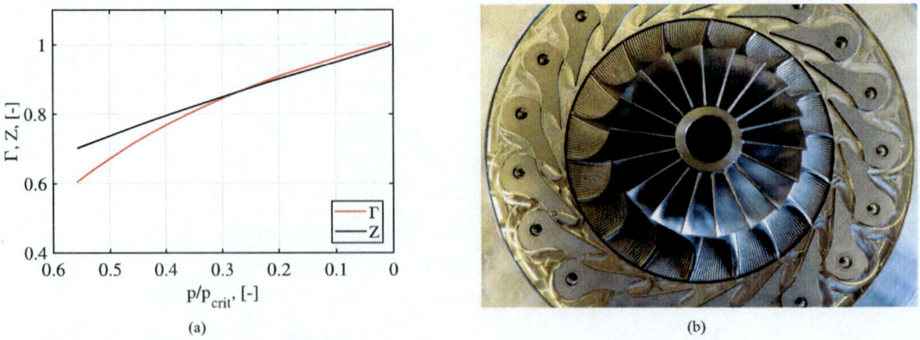

Fig. 1. a) Fundamental derivative of gas dynamics (Γ) and compressibility factor (Z) under isentropic expansion from turbine design inlet conditions to turbine design outlet pressure. b) Supersonic radial turbine of the experimental setup

system is to collect accurate experimental data on the operation and losses of small-scale high expansion ratio radial turbines for ORC applications.

Figure 1a shows the fundamental derivative of gas dynamics Γ and compressibility factor Z along the isentropic expansion from turbine inlet to outlet states at the design point. At the turbine design operating point, as well as in the typical operating conditions of the turbine, both the value of Γ and Z are well below 1 at the beginning of the expansion and the values are approaching unity at the turbine outlet. This indicates that the turbine is operated in non-ideal thermodynamic conditions, and the vapor expansion is highly non-ideal especially in the turbine stator. Non-ideal phenomena like the non-monotone development of Mach number along the expansion can also be encountered. An image of the turbine used in the experimental system is shown in Fig. 1b. The turbine has a supersonic stator with outlet Mach number of 2.4 and a radial turbine impeller.

2 Experimental Setup

2.1 Turbine Measurements

The turbine experiments with the original measurement system were carried out between 2016–2019 [5]. When analysing the test results it was observed that defining the turbine isentropic efficiency from the experimental data is sensitive especially on the accuracy of the temperature measurement at the turbine outlet. During the experimental campaign with the original measurement setup, the turbine efficiency was, on average, over 70% [5]. However, it was estimated that even a small deviation of 1–2 K in the turbine outlet temperature has a high impact (from about 3% to almost 6%) on the defined turbine isentropic efficiency, if compared to the estimated turbine efficiency in design conditions.

The sensitivity of turbine outlet temperature on turbine isentropic efficiency with turbine design inlet conditions and design outlet pressure is presented in

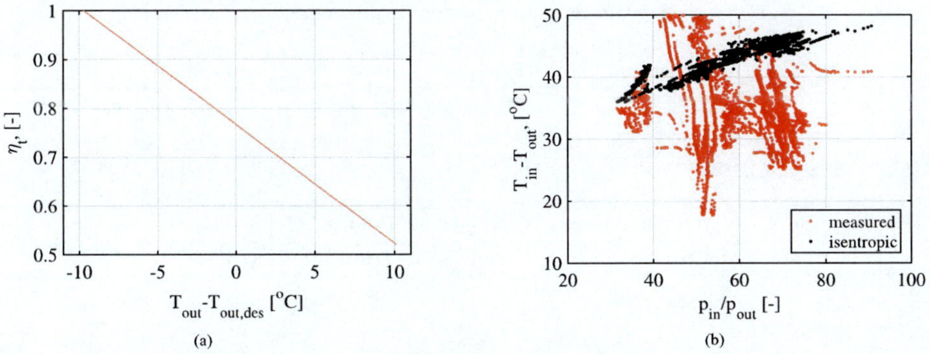

Fig. 2. (a) Sensitivity of turbine outlet temperature on turbine isentropic efficiency with turbine design inlet conditions and design outlet pressure. (b) Measured temperature drop over the turbine and isentropic enthalpy change defined from measured turbine inlet conditions and outlet pressure

Fig. 2a. During the experiments, the turbine outlet temperature measurement had some fluctuations resulting in high uncertainties in the turbine efficiency estimation defined from the measured data. Especially, these fluctuations in the measured turbine outlet temperature were observed to be more significant when running the turbogenerator at relatively low speeds, whereas with turbogenerator rotational speeds approaching the design speed, the measured turbine outlet temperature is more in line with the predictions of thermodynamic cycle models with less fluctuations during the operation [5]. An example of measured temperature drops over the turbine as a function of turbine pressure ratio and the isentropic temperature drop for the respective conditions are presented in Fig. 2b. The data shown in Fig. 2b are collected from measurement points of several test runs with the original measurement system containing data with different turbogenerator rotational speeds ranging from about 12 000 rpm to 30 000 rpm. The enthalpy drop over the turbine was based on the turbine inlet and outlet temperature measurements and the corresponding isentropic temperature drop was defined from the turbine inlet measurements and from turbine outlet pressure measurement. It can be observed that there is notable scattering of data especially at turbine pressure ratios between 40 to 60 and there are several measurement points where the measured temperature drop is even higher than the isentropic temperature drop at the respective conditions. On the other hand, with the higher pressure ratios between 60 to 80 there are less variations in the data and the measured temperature drops are at the expected range on average. It should be noted that in the majority of the data points with turbine pressure ratios between 60 to 80, the turbogenerator rotational speed has been relatively close to the turbine design rotational speed (from 24 000 rpm to 31 000 rpm), whereas the conditions with lower turbine pressure ratios are mainly representing operating conditions with turbogenerator rotational speeds below 23 000 rpm. The variations in the measured turbine outlet temperatures and in

Table 1. Measurement equipment

	Manufacturer	Type	Accuracy
Pressure	Gems	TR2200, [0–1,0–16 bar (abs)]	± 0.25% full scale
Temperature	Aplisens	CT-GN1	± 0.35 K (at 100 °C temperature)
Flow rate (liquid)	Kytola instruments	Oval gear SRP-40-H	± 0.5%
Flow rate (vapor)	McCrometer	V-cone VB0CAE01N	± 0.5%

data presented in Fig. 2b might be originating from both measurement accuracy and internal flows that might disturb the turbine outlet temperature measurement. One example of such internal flow could be a leakage of working fluid through the turbogenerator labyrinth seals to the turbine outlet, cooling down the turbine outlet temperature at some operating conditions. A more detailed analysis on the turbine performance under different operating conditions defined with the original measurement system and discussions on the potential sources of uncertainty in the turbine measurements are given in [5].

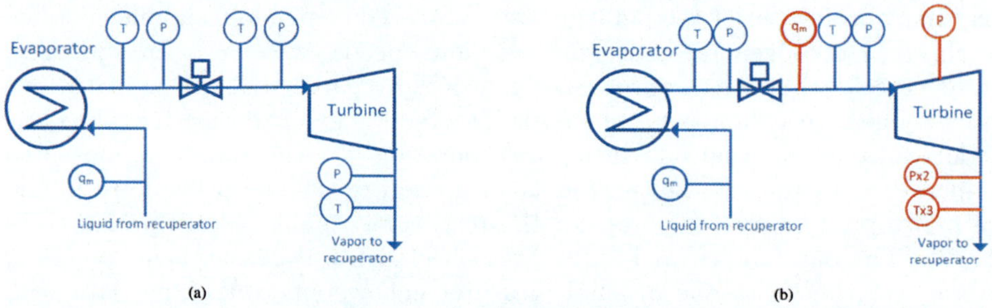

Fig. 3. (a) Original turbine related measurements and (b) updated turbine measurements. P = pressure measurement, T = Temperature measurement, q_m = flow measurement

The turbine measurement system has been updated to allow more accurate turbine measurements. The original measurement system is shown in Fig. 3a and the updated measurement system is shown in Fig. 3b. The information on the measurement equipment and the sensor types are given in Table 1. In the updated system, a V-cone sensor is added after the evaporator to measure the flow rate of the vapor working fluid entering the turbine. Besides increasing the reliability of the flow rate measurement, this second flow meter allows to detect the possible accumulation of working fluid in the evaporator. An oval gear meter is used for measuring the flow of liquid working fluid entering the evaporator. Two additional temperature sensors have been installed at turbine outlet to reduce the uncertainties in the turbine outlet condition measurements, since in

Fig. 4. Positioning of the temperature sensors at the turbine outlet duct

the old setup the turbine outlet temperature was measured by using only one sensor. The positioning of the three temperature sensors at the turbine outlet duct is presented in Fig. 4. The increased number and the placement of turbine outlet temperature sensors will help detecting the possible malfunctions in individual sensors as well as in detecting the possibility of uneven temperature distributions of the turbine discharge flow that can effect on the efficiency definition from the measurements. In addition, a second pressure probe was added at the turbine outlet to increase the reliability of the turbine outlet pressure measurement.

2.2 Pressure Measurements Inside the Turbine Stator

During a single experimental run, the operating conditions of the turbine experience significant variations. Pressure measurement points are added inside the turbine stator in order to monitor the expansion along the blade passages at varying operating point. Figure 1b shows the top view of the radial turbine of the experimental setup. The 19 blades of the stator ring delimit converging-diverging planar nozzles, with throat width 1.3 mm and blade height 2 mm. The stator has internal diameter 145 mm and external diameter 230 mm. Access to the flow is possible from the stator back plate visible in Fig. 1b. There, static-pressure wall tappings are machined at the positions highlighted in Fig. 5, and pressure is measured along the blade passages. This paragraph describes the measurement system and discusses the rationale behind the choice of measurement locations.

The most relevant constraint to implementing a measurement system inside the stator is represented by the small size of the stator ring. Indeed, conventional machining processes limit the minimum obtainable diameter of pressure taps to

Fig. 5. Drawing of the stator ring (to scale), equipped with pressure taps (center markings). Pressure measurement points are located in the convergent and in the divergent, both upstream and downstream of the trailing-edge shocks

0.5 mm, which compared to the characteristic lengths of the blade passages reads ∼1/3 of the throat width and 1/4 of the blade height. First, the spatial pressure gradient experienced by the expanding flow at the surface of pressure taps can be relevant due to the large relative size of wall tappings with respect to the blade passage. Second, cavity-like flow patterns are expected to occur past the exposed edges of pressure taps due to interaction with the main flow. Though the intrusiveness of the measurement system cannot be eliminated, mitigation of its effects can be achieved by carefully selecting the position of measurement points.

Results of detailed numerical analyses of the stator flow-field are examined in order to identify the regions characterized by low spatial pressure gradient. Reference results are those documented in Ref. [4], which include two-dimensional Navier-Stokes simulation of design and off-design operation of the stator. Further 2D Navier-Stokes simulations are performed in this study (Fig. 6) using ANSYS Fluent v19.4, RefProp v7.0 database, and k-ω SST turbulence model. Fig. 6a and 6b show the pressure distribution inside a stator channel at the design operating point (total pressure 7.9 bar, total temperature 538.15 K, and stator outlet pressure 0.4 bar). Figure 6a reports pressure isolines in the nozzle, while Fig. 6b shows the pressure (blue) and pressure gradient (red) profiles along the nozzle centreline, i.e. along yellow line in Fig. 6a. In Fig. 6b, the origin of the x axis is placed at the throat section. Results show that the expansion from inlet to outlet pressure levels occurs mostly in a 10 mm-long region close to the nozzle throat, where the pressure gradient reaches ∼3 bar/mm. Pressure experiences another sharp variation across the trailing-edge shock wave. Elsewhere, the pressure gradient is significantly smaller. Three radial locations are selected, one in the convergent, one in the divergent upstream of the trailing-edge shocks, and one at stator outlet past the trailing-edge shocks, see Fig. 5. At the selected locations, pressure gradient is < 20 mbar/mm, and the pressure variation across the surface of pressure taps is within the accuracy of pressure transducers.

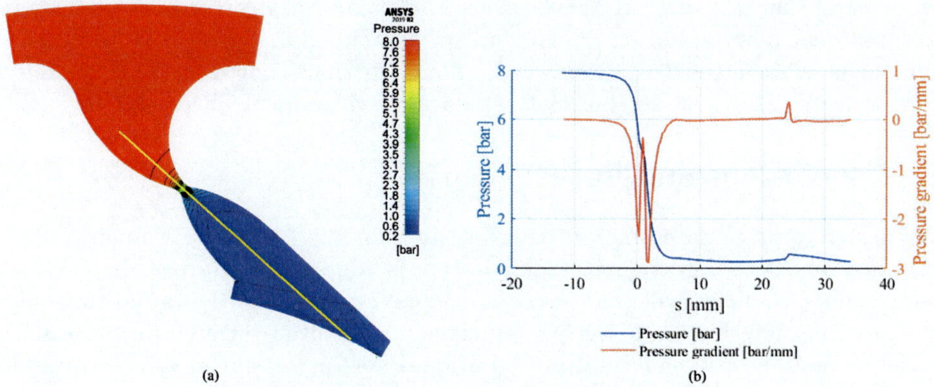

Fig. 6. (a) Pressure distribution within the stator channel in design conditions (b) Pressure and pressure gradient profiles along the yellow line of Fig. 6a

A quantitative analysis of cavity flows past the edges of wall tappings is not performed in this study. However, their occurrence is considered and its impact on the final pressure readings is analyzed. Given the low diameter/depth ratio of wall tappings, an open-cavity flow configuration is expected to occur, see e.g. Ref. [3], with flow separation at the upstream edge of the tapping, recirculation inside the cavity, and reattachment at the downstream edge. The incoming flow experiences a slight expansion at the upstream edge of the tapping, followed by a recompression at the downstream one. In the supersonic regions of the flow, these phenomena are accompanied by expansion and compression waves, which will propagate and perturb the main flow. Though the final pressure readings are not expected to depart significantly from the target pressure levels, it is unclear to which extent these perturbations will affect the main flow during the experiments. For this reason, all tappings are located relatively far from the blade profiles, and pressures upstream and downstream of trailing-edge shocks are measured in separate blade passages. In order to evaluate flow uniformity around the stator ring, and to check the validity of pressure readings at each radial location, three sets of measurement points are used at different angular positions (cf. Fig. 5).

Industrial transducers of the same model adopted for pressure measurements at turbine inlet and outlet are employed, see the first line of Table 1. Transducers with full scale 16 bar(abs) and 1 bar(abs) are used, respectively, in the subsonic and supersonic regions. The accuracy of transducers is within $\pm 0.25\%$ of full scale, while their response time is ~ 0.5 ms. Transducers are connected to the wall tappings through a pressure line. A linear analysis of the resulting pressure line-cavity system is carried out according to the formulation presented in Ref. [6], which yields an estimated maximum response time of about 10 ms.

Combined with pressure and temperature measurements at turbine inlet and outlet, pressure measurements inside the stator allow to quantify the contribution of stator and rotor to the overall turbine expansion, and to estimate trailing-edge shock losses. Moreover, since the expansion in the stator occurs in the non-ideal flow regime, a dependence of the expansion profile on the total conditions is expected to be observed, though perhaps not directly inferrable from measurements due to the relatively large measurement uncertainties.

3 Conclusions and Future Works

The design of an updated measurement system for the LUT 10 kW micro ORC test-rig is presented. Further instrumentation is adopted to improve the performance analysis of the cycle and its components, especially of the radial turbine. A better characterization of turbine efficiency is achieved through an increased sensor redundancy, namely with an additional flow meter ahead of the turbine and duplication of temperature and pressure probes at turbine outlet. Pressure measurement points are also added inside the turbine stator ring to characterize the supersonic expansion along the stator blade passages, to estimate trailing-edge shock losses, and to assess how the expansion is divided between the stator and rotor during the experiments.

The updated measurement system is currently being commissioned, and the first experimental campaign will be carried out in Autumn 2020. From the acquired data and the use of state-of-the-art thermodynamic models, detailed performance maps of the operation and losses of the system will be redacted, and the influence of flow non-ideality on the performance parameters will be investigated. Experimental data will also be used to validate the numerical models of the turbine flow-field in design and off-design conditions.

References

1. Colonna, P., et al.: Organic rankine cycle power systems: from the concept to current technology, applications, and an outlook to the future. J. Eng. Gas Turbines Power **137**(10) (2015). Article No. 100801
2. Seume, J.R., Peters, M., Kunte, H.: Design and test of a 10kw ORC supersonic turbine generator. J. Phys: Conf. Ser. **821**(1), 012023 (2017)
3. Stallings, R.L., Wilcox, F.J.: Experimental cavity pressure distributions at supersonic speeds. Technical report Technical Paper 2683, NASA (1987)
4. Uusitalo, A.: Working fluid selection and design of small-scale waste heat recovery systems based on Organic Rankine Cycles. Ph.D. thesis, Lappeenranta-Lahti University of Technology (2014)
5. Uusitalo, A., Turunen-Saaresti, T., Honkatukia, J., Dhanasegaran, R.: Experimental study of small scale and high expansion ratio ORC for recovering high temperature waste heat. Energy **208** (2020). Article No. 118321
6. VKI-LS-2001-04: Measurement techniques in fluid dynamics. VKI (2009)
7. Weiss, P.A., Popp, T., Müller, J., Hauer, J., Brüggemann, D., Preissinger, M.: Experimental characterization and comparison of an axial and a cantilever micro-turbine for small-scale Organic Rankine cycle. Appl. Therm. Eng. **140**, 235–244 (2018)

The Development of a Generic Working Fluid Approach for the Determination of Transonic Turbine Loss

John J. Otter(✉), David Baumgärtner, and Andrew P. S. Wheeler

University of Cambridge, Cambridge, UK
{jjo26,db701,aw329}@cam.ac.uk

Abstract. Due to the vast number of potential working fluids incorporated within heat recovery cycles, there is a degree of uncertainty around the aerodynamic design and performance of the required turbomachinery. This paper aims to explore the aerodynamic performance of transonic turbines in a generic manner so that the accurate prediction of turbine performance can be made for a wide variety of working fluids. Within this study, a thermodynamic approach to model a generic working fluid is presented based on the Peng-Robinson equation of state. This model is used in combination with Computational Fluid Dynamics to complete a systematic study into how various fluid parameters impact turbine aerodynamic performance. Therefore, this work presents a step towards a turbine loss model which is applicable to any transonic vapour turbine.

Keywords: Computational Fluid Dynamics · Organic Rankine Cycle · Loss coefficient

1 Introduction

In the preliminary design of turbines for heat recovery and Organic Rankine Cycles, correlations are often used to predict turbine performance [5,9]. A common set of such correlations which is often utilised are those reported by Aungier [2]. However, as these correlations are mainly derived from turbines operating with air, the validity of applying such correlations to turbines operating with fluids at non-ideal conditions has not been assessed. To this end, the aim of this study is to develop a generic approach to evaluate loss within transonic turbines such that the gas dynamics of the employed working fluid can be taken into account. This is achieved through the use of Computational Fluid Dynamics (CFD) to evaluate turbine performance for a series of parametric working fluids at multiple operating conditions. An array of working fluids are investigated through the use of the Peng-Robinson Equation of State (PREOS).

M. Pini et al. (Eds.): NICFD 2020, ERCOFTAC Series 28, pp. 123–131, 2021.
https://doi.org/10.1007/978-3-030-69306-0_13

2 Methodology

2.1 Thermodynamic Modelling

Within this study, the PREOS is used to model the working fluid (Eq. 1) from Ref. [6]. Although the PREOS is known to have deficiencies, especially around the fluid critical point, it was chosen for this study based on its relative simplicity compared to more complex multi parameter models such as those developed by Colonna et al. [4]. Moreover, the aim of this study is not to accurately model existing fluids, but instead to study how loss generation depends on non-dimensional parameters. Simplicity in this context applies to the implementation within the CFD solver and in terms of minimising thermodynamic degrees of freedom. The approach to studying a parametric fluid with the PREOS is based upon the work of White et al. [9] who studied the thermodynamic performance of an Organic Rankine Cycle operating with a generic working fluid.

$$p = \frac{RT}{V_\mathrm{m} - b} - \frac{a\alpha(T)}{V_\mathrm{m}^2 + 2bV_\mathrm{m} - b^2} \tag{1}$$

For brevity, a full description of the variables in Eq. 1 is not given here: only those pertinent to this study are discussed. The reader is instead referred to Refs. [6,9]. In order to complete the thermodynamic model, the ideal heat capacity c_p^{ideal} as a function of temperature is specified as a second order polynomial, $c_p^{\mathrm{ideal}} = A + BT + CT^2$.

With this model, the enthalpy and entropy of a fluid can be evaluated with the PREOS departure functions [7] and through the ideal gas contribution. The fluid saturation dome can be obtained through the use of Maxwell's rule [10] and the aforementioned departure functions. The PREOS places no constraints on the transport properties of the working fluid and hence these must be defined to undertake any CFD simulations. The dynamic viscosity of the fluid was modelled as a constant, i.e. not a function of temperature. The value of the dynamic viscosity μ was varied such that the mass-averaged Reynolds number at the exit of the turbine matched a pre-defined value which in this case was 1.7 million. The specific choice of Reynolds number is not of interest in the current context as the effect of Reynolds number on turbine loss is a well-researched topic. The thermal conductivity was set to achieve a molecular Prandtl number of 0.7 based on the mass-averaged specific heat constant at the exit of the turbine domain.

2.2 Non-dimensional Parameters

Two non-dimensional parameters are considered within this study, the compressiblilty factor at the turbine inlet $Z_0 = \frac{p_0}{\rho R_\mathrm{s} T_0}$ and the isentropic exponent $k = \log\left(\frac{p}{p_0}\right) / \log\left(\frac{\rho}{\rho_0}\right)$. Previous studies have shown that k has a large impact on transonic turbine loss [3] but the effect of Z_0 on loss is less well understood.

For this study the compressibility factor is evaluated based on the turbine inlet conditions; however the value of compressibility factor will vary over the fluid expansion through the turbine. The value of isentropic exponent is evaluated through the assumption that the fluid behaves in a polytropic manner across the expansion. As per the work of Wheeler et al. [8], a linear regression of the logarithms of pressure and density over the isentropic expansion process was used to determine k. As such, this can be interpreted as an average value throughout the expansion.

2.3 Computational Method

The aerodynamic performance of each turbine was evaluated through the solution of the 2D Reynolds-averaged Navier-Stokes equations. As the approach is 2D, the effect of the working fluid on secondary loss mechanisms is not considered. However, the approach outlined in this paper could easily be expanded to study such loss mechanisms. The implicit density-based solver from the commercial package ANSYS FLUENT V17.0 was used with the working fluid modelled using the inbuilt PREOS functionality. Closure of the RANS equations was achieved with the Spalat-Allmaras model. Computational meshes were generated with ANSYS Turbogrid and consisted of 100,000 nodes. The wall distance for each mesh was set to give a y^+ of 30 such that wall functions could be used. This computational approach has been validated and verified by Baumgärtner et al. [3].

2.4 Geometry Generation

The methodology for the generation of the turbine blade geometries has been previously reported by Baumgärtner et al. [3] and hence is not reported in detail hereafter. All geometries were modelled as part of a rectilinear cascade with an outlet angle of 75°. For supersonic outflow conditions ($M_{\text{exit}} > 1.1$) the turbine geometries were created based on the specific value of k and exit Mach number M_{exit}. With this approach, the method-of-characteristics was used to generate a geometry to expand the flow to a given value of M_{exit}. An example set of geometries is shown in Fig. 1a for various values of k. From this figure it can be seen that for a constant M_{exit}, lower values of k result in longer blades relative to higher values of k. For the transonic exit case ($M < 1.1$) a constant geometry was used based on a common transonic blade geometry [1]. In the current methodology, there is no change in geometry for different values of Z_0.

3 Results

3.1 Thermodynamic Design Space

The aim of the section is to define a thermodynamic design space in which Z_0 and k can be varied independently of each other. This is desirable as it allows the impact of each variable on turbine loss to be determined independently. However, a full decoupling of these variables is not possible for a single working

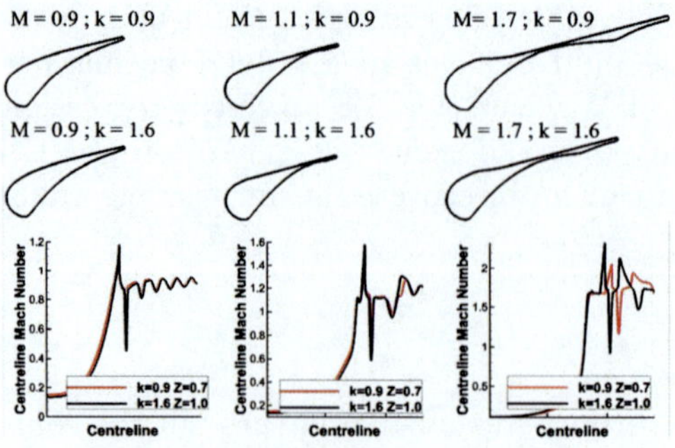

(a) *Example family of blades for* $k = 0.8$ *and* $k = 1.6$

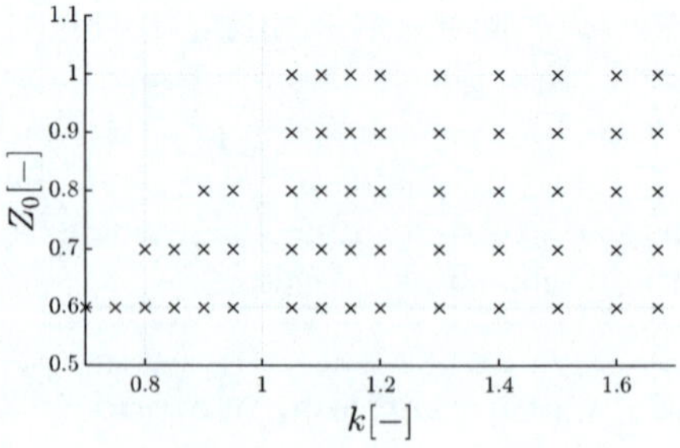

(b) *Range of k versus* Z_0 *at* $M_{\text{exit}} = 1.1$

Fig. 1. Geometric and thermodynamic design space

fluid as Z_0 and k are thermodynamically linked. Therefore, in order to complete a parametric study of compressibility factors and isentropic exponents, an array of working fluids must be defined.

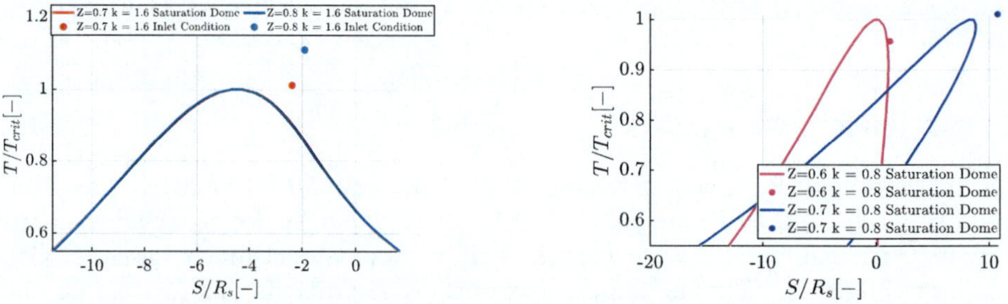

Fig. 2. Example T-s diagrams

The first step in the process is to set the desired value of M_{exit} as this sets the exit pressure and the effective blade loading. The second and third steps are to specify the target inlet compressibility factor and isentropic exponent. A variety of target values were investigated in this work to cover a range of potential turbine operating conditions. The desired range of Z_0 ranged from 0.6 to 1.0, k from 0.8 to 1.6 and M_{exit} from 0.8 to 1.7. An example design space for $M_{\text{exit}} = 1.1$ is shown in Fig. 1b. Within this figure, each cross corresponds to a different working fluid, such that in this example there are a total of 58 different fluids. This example highlights that is was not possible to fully cover the design space as values of k less than 1 cannot be reached unless Z_0 is below 0.8.

The procedure to obtain a given value of k and Z_0 for a given M_{exit} is outlined below. Firstly, the accentric factor and molar mass must be chosen. Secondly, an initial stagnation pressure is set based on a given reduced pressure. With p_0 fixed, Z_0 is only a function of temperature, and hence the stagnation temperature is given by $T_0 = \frac{p_0}{\rho R_s Z_0}$. The procedure to match k is more involved as both k and the turbine exit pressure depend on the polynomial coefficients that define c_p^{ideal} An initial guess for the values of A, B and C is used to determine the exit pressure which gives the desired value of M_{exit}. For this expansion, the value of k is the calculated; then an iterative procedure is used to find the values of A, B and C which minimises the difference between the calculated value of k and the desired value. Due to the non-linear nature of the design space, a global optimisation algorithm, in this case NSGA-II, was found to be the most reliable method to find a solution.

A set of example T-s diagrams for four different working fluids is presented in Fig. 2. Within this figure, the saturation domes are shown by the solid line and point markers indicate turbine inlet conditions. In this example, the saturation dome for two so-called wet fluids is shown in the left figure and two operating points are shown with symbols. These symbols correspond to inlet conditions with various values of Z_0. This example indicates how Z_0 can be varied by changing inlet temperature as the saturation dome for the two fluids is very similar. In addition, the saturation dome for two retrograde, or dry, fluids are shown in the right hand figure. These two cases feature a constant value of $k = 0.8$ and two values of Z_0.

As the PREOS is used, fluids are able to expand into the saturation dome as during the CFD simulations the vapour value from the equation of state is taken. This is only valid up until the spinodal point and cases where the turbine expansion crosses the spinodal point are not considered as the equation of state is no longer stable. Another two constraints on operating point are that the turbine inlet conditions cannot start within the saturation dome and that cases are only considered with $s > s_{\text{crit}}$.

3.2 Variation of Loss Coefficient with Isentropic Exponent and Compressibility Factor

Within this section, the sensitivity of loss coefficient to k and Z_0 is assessed. The loss coefficient was defined based upon the difference between the mixed-out enthalpy and isentropic enthalpy, $\zeta = (h - h_s)/(h_0 - h_s)$. In this context h denotes the mixed-out enthalpy derived from the CFD, h_0 denotes the local stagnation enthalpy and h_s represents the enthalpy that would be obtained if the fluid were to be isentropically expanded to a uniform pressure. The mixed-out enthalpy for both values of enthalpy were obtained from a constant area mixing calculation.

All the results presented were evaluated with inlet conditions corresponding to a reduced pressure of 0.7 and each fluid had a molar mass of 72 and an accentric factor of 0.251. The values of molar mass and accentric factor were chosen based on the corresponding values for pentane.

The variation of loss coefficient with k at exit Mach numbers of 0.9 and 1.3 are given in Figs. 3a and 3b respectively. For both exit Mach numbers, ζ is a monotonic function of k with Z_0 equal to 1 and 0.8. The relationship between k and ζ is linear; this is shown by the regression curves which both have coefficients

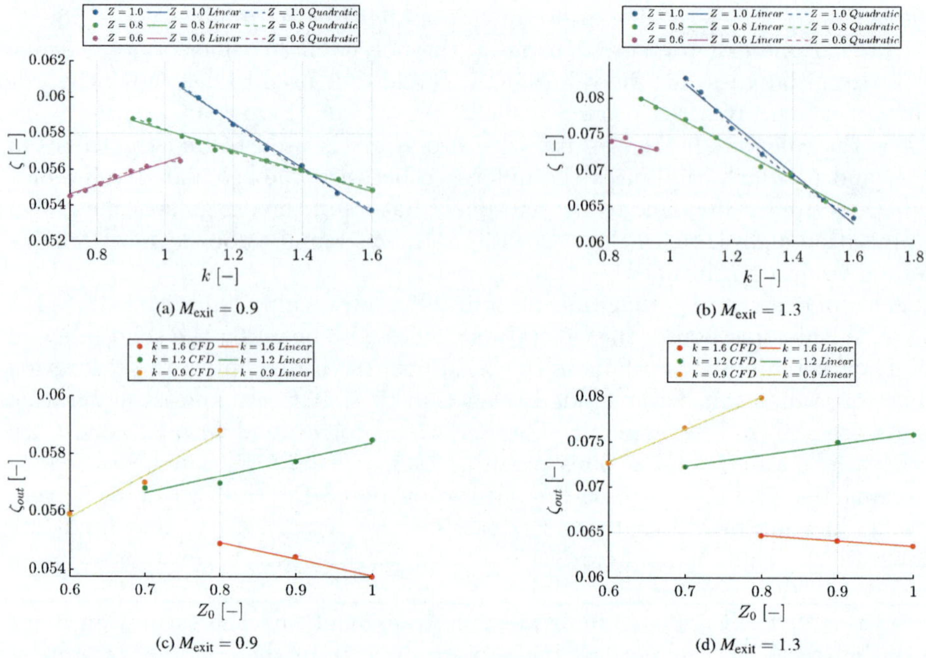

Fig. 3. Variation of turbine loss with isentropic exponent and compressibility factor

of determination that are greater than 0.99. This linear gradient $\frac{\partial \zeta}{\partial k}|_{Z_0}$ decreases as Z_0 is reduced. Therefore, the dependency of loss coefficient with k decreases at lower values of Z_0. However, for the case of $M_{\text{exit}} = 0.9$ and $Z_0 = 0.6$ the loss coefficient follows a non-linear relationship with k and the loss coefficient can be seen to decrease with reduced k (purple line in Fig. 3a). At $M_{\text{exit}} = 1.3$ and $Z_0 = 0.6$, the loss coefficient increases with lower values of k but no longer behaves in a linear manner (purple line in Fig. 3b).

The variation of loss coefficient with Z_0 for two exit Mach numbers is shown in Figs. 3c and 3d. The first point to make from these figures is that a greater variation in ζ is observed over the range of k values as compared to the variation with Z_0. For both exit Mach numbers, the qualitative behaviour of loss with Z_0 was found to be the same. That is at $k = 1.6$, the gradient $\frac{\partial \zeta}{\partial Z_0}|_k$ is positive where as at $k = 1.2$ and 0.9 this gradient becomes negative and increases in magnitude, so the impact of Z_0 is greater for lower values of k. Another point of interest is that the linear gradient $\frac{\partial \zeta}{\partial Z_0}|_k$ peaks at an exit Mach number of 1.5 for k values of 0.9, 1.2 and 1.4.

In order to contextualise these variations in loss coefficient, they can be normalised by the corresponding loss coefficient of air. To this end, Fig. 4 shows four maps of this loss coefficient ratio as a function of Z_0 and k. At $M_{\text{exit}} = 0.9$ the variation is fairly modest with a maximum 5% increase in ζ occurring at $Z_0 = 1$ and $k = 1.05$ (Fig. 4a). At $M_{\text{exit}} = 1.7$, the variation is at its highest

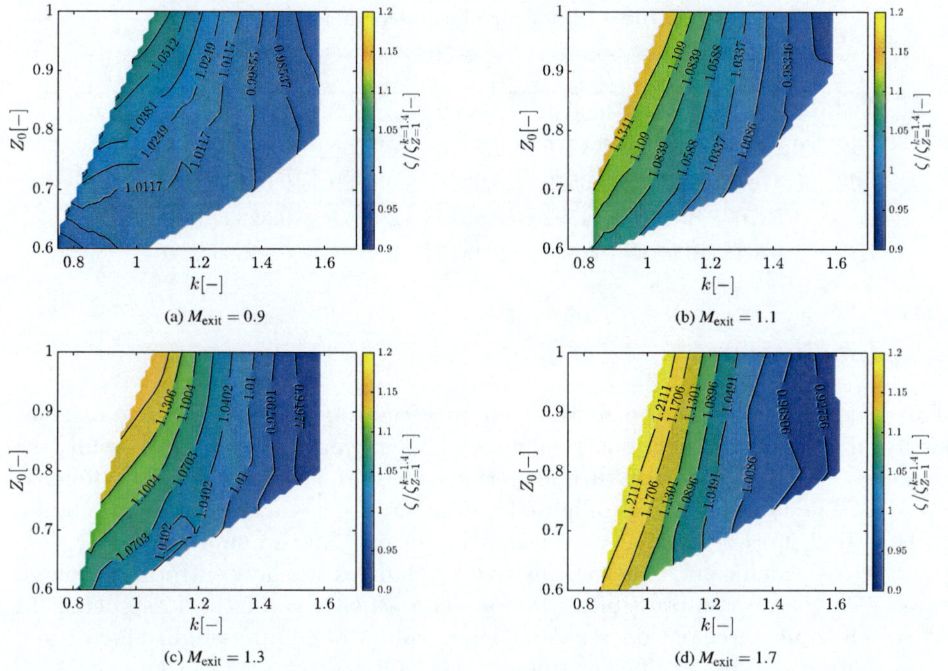

Fig. 4. Variation of turbine loss with compressibility factor, isentropic exponent and exit mach number

with a 20% increase in loss coefficient occurring at the lowest k values for all Z_0 (Fig. 4d). In all cases with supersonic exit conditions, the lowest value of loss occurred at $k = 1.6$ for Z_0 values of 0.8 to 1.0. Note that not all points on the map produced a converged CFD solution for each exit Mach number and hence there is not an equal number of points in each map. Most notably convergence problems occurred at lower values of Z_0 and k.

Based on these results, it is possible to construct a preliminary correlation which accounts for variation to k and Z_0 to the loss coefficient. For simplicity, a two-dimensional quadratic formulation was chosen (Eq. 2). The coefficients for this correlation are given in Table 1 for each value of M_{exit} as well as values of the root mean squared error. Given the values of this root mean squared difference vary from 1 to 3% this correlation would be suitable to obtain an initial estimate for loss variations. Care must be taken to ensure that results are not taken from outside of the explored $k - Z_0$ range as indicated in Fig. 4. Note that no coefficients are reported at $M_{\text{exit}} = 0.8$ as the variation in the loss ratio was less than the root mean squared error.

$$\zeta/\zeta^{k=1.4}_{Z_0=1} = \alpha_{00} + \alpha_{10}k + \alpha_{01}Z_0 + \alpha_{20}k^2 + \alpha_{11}kZ_0 + \alpha_{02}Z_0^2 \tag{2}$$

Table 1. Loss correlation coefficients

M_{exit}	α_{00}	α_{10}	α_{01}	α_{20}	α_{11}	α_{02}	$RMSD(\%)$
0.90	0.5051	0.4714	0.6065	−0.0069	−0.6828	0.1954	3.24
1.10	0.7813	−0.1614	1.1476	0.2228	−0.8286	0.0205	0.55
1.30	0.4768	0.2202	1.3579	0.0958	−0.9809	0.0445	0.95
1.50	1.5276	−1.3873	1.4562	0.6614	−0.9341	−0.0290	1.18
1.70	1.3762	−0.9020	1.1858	0.4675	−0.9728	0.1339	2.00

4 Conclusions

This study has presented a method for investigating the performance of transonic turbines that is not constrained to a known working fluid. As a result, the impact of fluid properties on turbine performance can be explored in a systematic manner. The initial study highlighted a number of generalised points in relation to how fluid properties affect performance. For exit Mach numbers greater than 0.9, the loss coefficient is more sensitive to changes in the isentropic exponent than it is to the compressibility factor. The sensitivity of the loss coefficient with isentropic exponent decreases at lower values of compressibility factor and the sensitivity of loss coefficient to compressibility factor peaks at an exit Mach number of 1.5. An example of how this methodology can be used in a preliminary turbine design process has been demonstrated through the formation of a correlation which can account for variations in working fluids on turbine loss. Future work will look to investigate other fluid properties such as the accentric factor as well as exploring a larger thermodynamic design space.

Acknowledgments. This work was supported by the EPSRC (EP/L027437/1) and computational resources were provided by EPSRC Tier-2 capital grant EP/P020259/1. The authors would like to thank the reviewers for their insightful comments and valuable feedback.

References

1. Arts, T., de Rouvriot, M.L., Rutherford, A.W.: Aero-thermal investigation of a highly loaded transonic linear turbine guide vane cascade: A test case for inviscid and viscous flow computations. Technical report, Von Karman Institute Technical Note 174 (1990)
2. Aungier, R.H.: Turbine Aerodynamics: Axial-Flow and Radial-Flow Turbine Design and Analysis. ASME Press, New York (2006)
3. Baumgärtner, D., Otter, J.J., Wheeler, A.P.S.: The effect of isentropic exponent on transonic turbine performance. J. Turbomach. **142**(8), 081007 (2020)
4. Colonna, P., Nannan, N., Guardone, A., Lemmon, E.: Multiparameter equations of state for selected siloxanes. Fluid Phase Equilib. **244**(2), 193–211 (2006)
5. Da Lio, L., Manente, G., Lazzaretto, A.: A mean-line model to predict the design efficiency of radial inflow turbines in organic Rankine cycle (ORC) systems. Appl. Energy **205**, 187–209 (2017)

6. Peng, D.Y., Robinson, D.B.: A new two-constant equation of state. Ind. Eng. Chem. Fundam. **15**, 59–64 (1976)
7. Sandler, S.I.: Chemical, Biochemical and Engineering Thermodynamics. Wiley, Hoboken (2017)
8. Wheeler, A., Ong, J.: The role of dense gas dynamics on ORC turbine performance. J. Eng. Gas Turbines Power 135, 102603 (2013)
9. White, M.T., Sayma, A.I.: A generalised assessment of working fluids and radial turbines for non-recuperated subcritical organic Rankine cycles. Energies **11**(4), 800 (2018)
10. Wisniak, J., Golden, M.: Prediction of the saturation curve of a pure substance using Maxwell's rule. J. Chem. Educ. **75**(2), 200 (1998)

Experiments and Measurement Techniques

Application of Hot-Wire Anemometry in the High Subsonic Organic Vapor Flow Regime

Felix Reinker(✉) and Stefan aus der Wiesche

Muenster University of Applied Sciences, 48565 Steinfurt, Germany
{f.reinker,wiesche}@fh-muenster.de

Abstract. Hot-wire anemometry (HWA) is a measurement technique which allows for the investigation of highly fluctuating flows. It has been widely used for turbulence studies including turbomachinery flow analysis, but relatively little is reported regarding its application in organic vapor flows. In this contribution, preliminary results are presented of an HWA study conducted in the test section of a closed-loop organic vapor wind tunnel using a constant-temperature-anemometry (CTA) system. The working fluid for this study was the perfluorinated ketone Novec™ 649, but air was also investigated for comparison reasons. Low subsonic up to high subsonic flows (i.e., up to $M = 0.7$) were considered, and the performance of HWA was assessed by obtaining turbulence intensities, velocity spectra, autocorrelation, and corresponding turbulent length scales. It was demonstrated that HWA is a useful tool for investigating turbulence phenomena in organic vapor flows although the dynamic loads and the wire Reynolds number levels are rather high.

Keywords: Experimental techniques · Hot-wire-anemometry · CTA · Turbulence · Organic vapor flows

1 Introduction and Scope of Research

The application of computational fluid dynamics (CFD) tools for the design and optimization of organic Rankine cycle (ORC) turbines has become a relevant topic in terms of sustainable power systems, and it is still growing (Colonna et al. 2015; Macchi and Astolfi 2017). However, there is a real need for experimental investigations and reliable flow measuring techniques in order to validate CFD methods for modelling and simulation of non-ideal gas flows.

The flow within a turbomachine is in general unsteady and turbulent. Hot-wire anemometry (HWA) is a measurement technique which allows for the investigation of highly fluctuating flow phenomena (Smolyakov and Tkachenko 1983), and it is still the primary tool in turbulence research, but relatively little is reported about its performance in organic vapor flows. So far, compressibility effects were investigated mainly under ideal gas conditions. HWA is a thermal measurement technique, based on the convective heat transfer between a heated thin cylinder (wire) and the relevant fluid (Bruun 1995). Because of the very fast thermal response of the thin wire, HWA allows

M. Pini et al. (Eds.): NICFD 2020, ERCOFTAC Series 28, pp. 135–143, 2021.
https://doi.org/10.1007/978-3-030-69306-0_14

for the investigation of highly fluctuating flow velocities. The scientific application of HWA has a long history. Although HWA is mostly used for obtaining velocities, it is from a physical point of view rather an area-averaged mass flow measurement. HWA is strongly affected by compressibility effects and fluctuations of thermodynamic quantities particularly in transonic flows (Kovasznay 1950; Morkovin 1956; Horstman and Rose 1977; Souza and Tavoularis 1999). Another well-known challenge arising in high-speed flow applications arises from the high dynamic wire loads, increasing the risk of wire breakage and short life time of the probes. Hence one might expect serious problems for HWA in organic vapor flows, because of the high density levels due to their high molecular complexities (although the low speed of sound might lead to somewhat moderate absolute flow velocities). A statement on the actual HWA breakage risk in organic vapor flows can only be given by means of experiments performed under realistic flow conditions. Apart from these issues, contamination and corrosion of the wire are also crucial because both can strongly disturb calibration efforts. Corrosion might be relevant because the hot-wire material is typically not stainless steel. Due to the fact that organic vapor test facilities and power systems are always characterized by a closed hermetic design, contamination with air dust should not be a serious issue.

Currently, there are only few operational wind tunnel test facilities available for investigating organic vapor flows. One of these test facilities is the closed-loop organic vapor wind tunnel (CLOWT), built up at Muenster University of Applied Sciences (Reinker et al. 2018; Reinker et al. 2019). Since CLOWT has been designed as a continuously running wind tunnel with the possibility to calibrate probes, it is a powerful tool for assessing the potential of HWA in the high subsonic flow regime. In this contribution, the outcome and the experiences are presented of a first preliminary experimental study conducted at CLOWT using a constant-temperature-anemometry (CTA) system.

2 Experimental Setup and Procedure

The considered CTA system was a modified system designed for higher wire temperatures with an overheat ratio of about $\tau \approx 2.3$ using a robust heated wire (diameter: $d = 10$ μm; length: $L = 4$ mm; material: gold-plated tungsten) provided by SVMtec GmbH, Stuttgart. Figure 1 shows the corresponding microscopic photograph and the dimensions of the CTA probe utilized in this study. The present experiments were performed in the calibration test section of the closed-loop organic vapor wind tunnel (CLOWT) using the perfluorinated ketone Novec™ 649 and dry air (see Fig. 2c). More details on the test facility CLOWT and its working principle can be found in previous publications (Reinker et al. 2018; Reinker et al. 2019).

After passing the centrifugal compressor, the working fluid was decelerated in the diffuser and entered the settling chamber where stagnation pressure p_0 (uncertainty $\Delta p/p = 0.7\%$) and total temperature T_0 (uncertainty $\Delta T/T = 0.18\%$) were measured. The mass flow rate m was measured by means of a mass flow device in the return of the wind tunnel (with maximum uncertainty level of $\Delta \dot{m}/\dot{m} = 1.6 - 4.5\%$ depending on the actual operation conditions). Different turbulence intensities could be provided by means of variable screen sets, but during the present study, the settling chamber was not equipped with additional flow straighteners or screens in order to provide noticeable turbulence intensities (of order 0.5 up to 1.0%). In CLOWT, the flow is successively accelerated in a two-stage contraction zone. The first subsonic axisymmetric nozzle (from standard diameter DN500 to DN250) offers a moderate contraction ratio of about 3.7, whereas the second nozzle accelerates the flow to high subsonic or transonic flow, depending on the chosen modular nozzle device (see Fig. 2a). In the present study, the second contraction was achieved by a three-dimensional nozzle, based on additive manufacturing, providing a three-dimensional cross-section change from round to rectangular contraction (DN250 to 50 mm $\times$ 100 mm) leading to a total contraction ratio of about 38 (see Fig. 2c on the right). The probe blockage of 2% in the calibration test section was negligible in terms of disturbances on the static pressure p_1 which was measured at endwall pressure taps ahead of the hot wire (see Fig. 2b).

Actual vapor states and Mach numbers M in the test section were calculated using REFPROP data based on recorded pressures, temperatures, and mass flow rates and assuming isentropic nozzle flow. The heat transfer at the hot wire was driven by the temperature difference $T_w - T_f$ between hot wire and fluid. Special attention required the determination of the correct value of the wire temperature, T_w. For this purpose, the probe and electrical line resistances R_L were determined and then the wire temperature, T_w, which is a function of the temperature-resistance dependency of the wire material, was calculated. Details on this method can be found in (Bruun 1995). To calculate the fluid temperature T_f, which in fact was the recovery temperature, the recovery factor was assumed to r = 0.89, based on the Prandtl number of Novec™ 649. For signal processing a highspeed analog input module NI9229 of National Instruments (Delta-Sigma digitizer and analog prefiltering) was used with a sampling rate of 10 kHz and sampling duration of t = 4s. The relative uncertainty of the electrical bridge voltage $\Delta U_B/U_B$ was of order 1.5 up to 2.0%.

In a first set of experiments, a static calibration of the HWA system with Novec™ 649 as working fluid was performed at different densities (up to ρ = 35 kg/m^3), pressure levels (up to 3 bar), but at a fixed total temperature of about T_0 = 95 °C $\pm$ 0.5 °C. The mass flow rate was controlled by variation of the compressors running speed, n, leading to a Mach number range of about M = 0.1 up to 0.7. A laminar heat transfer correlation with coefficient functions A and B was assumed for the static calibration. In Eq. (1) U_B is the Wheatstone bridge voltage of the HWA, w is the ambient flow velocity, based on mass flow rate and cross-sectional area of the test section, d is the hot wire diameter, and μ is the viscosity.

$$\mathrm{Nu} = A(M,\tau) + B(M,\tau)\ \mathrm{Re}^{1/2} \quad \text{with} \quad \mathrm{Nu} \sim U_B^2/\left(T_w - T_f\right) \quad \text{and} \quad \mathrm{Re} = w\rho d/\mu \tag{1}$$

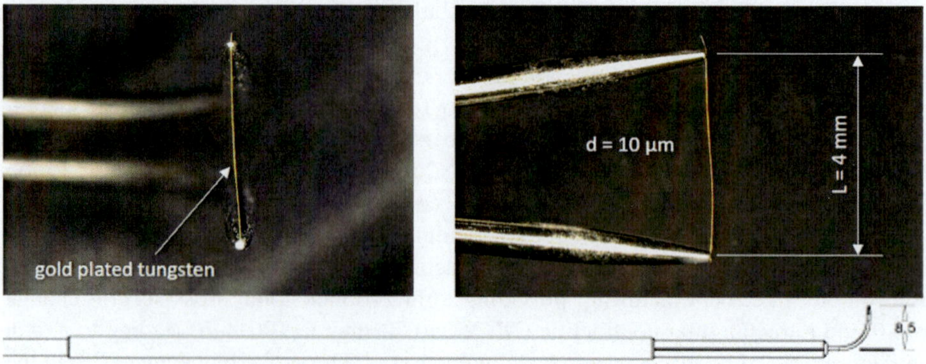

Fig. 1. Microscopic photograph and dimensions of the gold-plated tungsten hot-wire and the prongs.

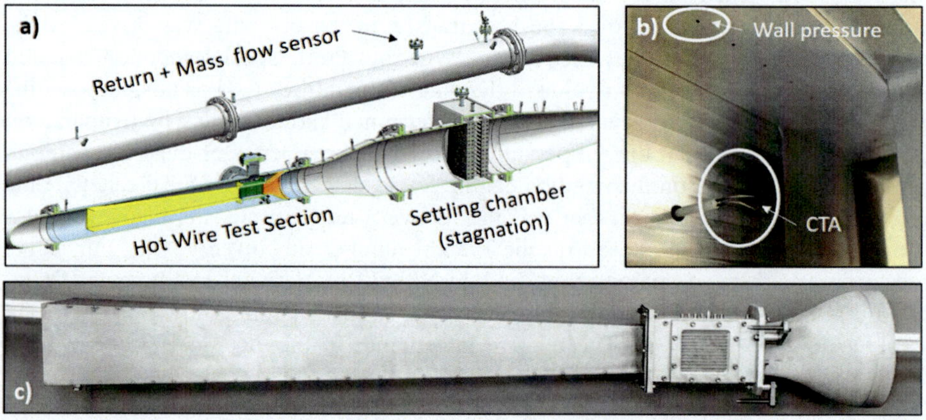

Fig. 2. a) CAD model of CLOWT with diffuser and settling chamber (stagnation conditions) on the right and hot wire test section on the left. b) Look into the test section including CTA mounting and wall pressure tap. c) Photograph of the modular high-speed test section including 3D-printed nozzle, rectangular modular test section, and diffuser (from right to left).

In a second step, the data from the calibration procedure was used to calculate the main flow velocities from the collected bridge voltage U_B. Using this velocity data set the turbulence intensity Tu was calculated by the relation $Tu = w'/\overline{w}$, with the root-mean-square of the turbulent velocity fluctuations w' and the mean velocity $\overline{w}$. Based on the assumption of isotropic turbulence, the formula considers only the main flow direction, which is obviously a simplification.

However, as an independent test of the turbulence level prediction, the classical turbulence sphere experiment, which is traditionally used for the characterization of low speed wind tunnels (Barlow et al. 1999), was conducted in the low speed test section (after the first contraction zone) at essentially incompressible flow conditions prior to the high-speed CTA measurements. Figure 3 shows on the left the polished turbulence sphere which was mounted in the centerline of the low speed test section. The corresponding drawing with global dimensions and positions of the pressure taps is shown on the right of Fig. 3. The stagnation pressure was measured at the leading edge and the base pressure at the aft surface, which was averaged by terms of four pressure taps. More details on the approved design of turbulence spheres and the corresponding calculation methods for the determination of Tu can be also found in (Barlow et al. 1999).

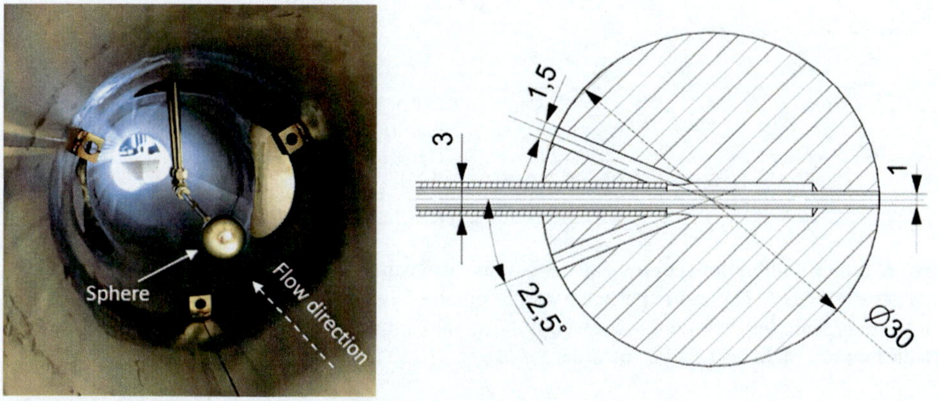

Fig. 3. Turbulence sphere in the basic test section (DN250) of the wind tunnel (left) and drawing of the turbulence sphere with all dimension (right). Front and rear pressure taps are equipped with separate pressure lines to measure the differential pressure.

3 Results and Discussion

The results of the static CTA calibration are shown in Fig. 4. It was found that the laminar heat transfer correlation (1) led to deviations less than 1.0%, and the typical deviation level was of only 0.3%. Inspection of Fig. 4 indicates that a systematic error trend was not obvious, but it should be noted that the highest deviation were observed at the highest Mach number (M = 0.7). Despite significant changes in density and other thermophysical material properties at other operating points, the laminar heat transfer correlation (1) still proofed its reliability.

The outcome of the turbulence investigations is shown in Fig. 5. On the left diagram, the results of CTA and turbulence sphere experiments conducted inside of the low speed test section under essentially incompressible flow conditions are shown. There was a good agreement observed between the CTA results and the independent turbulence sphere test. Interestingly, the flow control by the throttle valve in the return of the wind tunnel, see Fig. 2, led to higher turbulence intensities than the compressor running speed approach at fully opened valve. The effect of the second contraction on the turbulence intensity can be seen in Fig. 5 on the right. The turbulence intensity was reduced from its entry value of about 1.6% to a final level of about 0.5% after passing the second high-speed nozzle. It was observed that Tu increased systematically with increasing Mach number M to values of about Tu = 1.0% at M = 0.7.

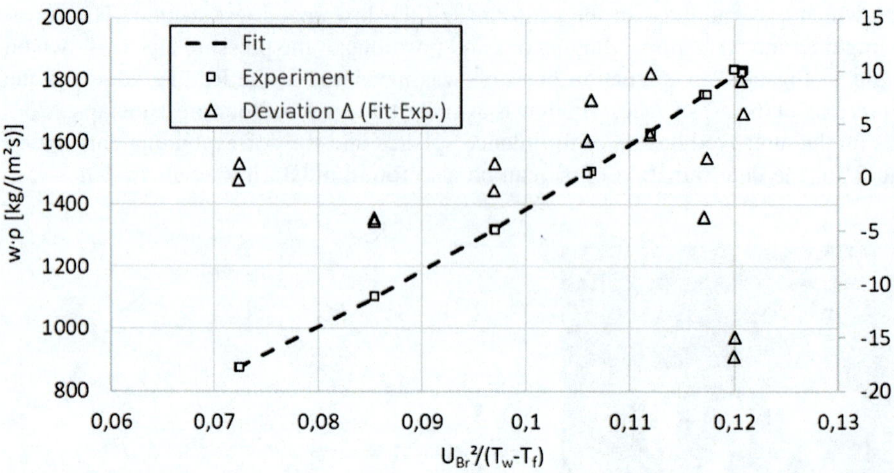

Fig. 4. Static calibration curve and deviations between fitting curve based on laminar heat transfer correlation (1) and experimental data obtained independently by mass flow sensor for subsonic organic vapor flow (Novec™ 649). The maximum error, which occurred at the highest Mach number of M = 0.7, was of order 0.9%.

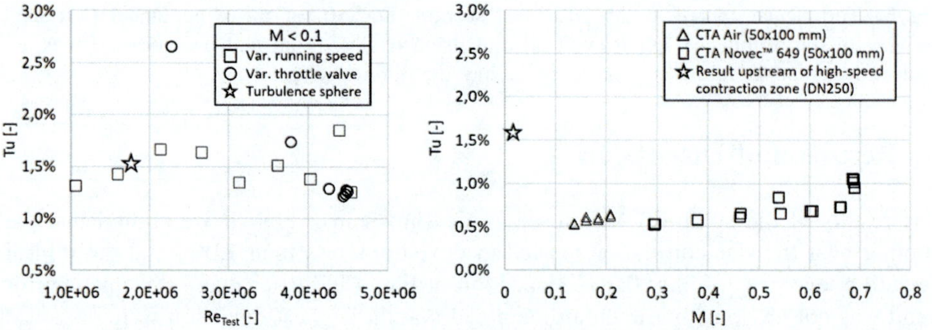

Fig. 5. Left: Turbulence intensities for Novec™ 649 in the low speed test section at different Reynolds numbers, measured by CTA and turbulence sphere. Right: Turbulence intensities, measured by CTA, for Air and Novec™ 649 in the high-speed test section downstream of the second contraction.

The spectral analysis of the CTA signals provided further insights into the turbulent flow behavior, but it exhibited also the impact of electrical noise due to the compressor drive and control unit. Figure 6 shows an example of such an un-filtered normalized power density spectrum (PSD) of the CTA signal U_B obtained at $M = 0.7$, corresponding to a compressor running speed of 3000 rpm (50 Hz).

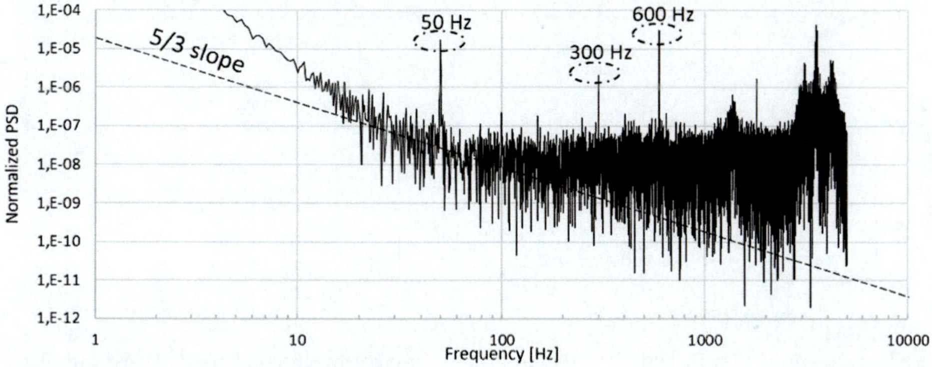

Fig. 6. Spectral analysis of an unfiltered (no cutoff filter) Hot-wire signal (Novec™ 649 at $\rho = 35$ kg/m^3 and M = 0.7) at 50 Hz compressor rotational speed. The rotor and blade passage frequencies are highlighted. The broken line corresponds to Kolmogorov's 5/3-law.

There are three regular or synchronous peaks observable in the power spectral density in Fig. 6, namely the rotational speed of the compressor at 50 Hz, the passage frequency of the six main compressor blades at 300 Hz, and the passage frequency of the twelve main and splitter blades at 600 Hz. The linear relation between compressor running speed n and peak frequencies was also observed for other rotational speeds much lower than 50 Hz (the effect shown in Fig. 6 was not caused by the European 50 Hz power grid and its harmonics, but is was related to the design of the compressor impeller). However, the high sensitivity of the employed CTA system was also demonstrated by the clear resolution of these synchronous peaks. Additional screens in the settling chamber would noticeably reduce these regular patterns, but for the present purpose, they served as benchmark.

The increasing deviation from Kolmogorov's 5/3-law at higher frequencies became obvious in Fig. 6, indicating the necessity to remove the impact of external electrical noise by terms of low-pass filtering of the CTA signal. The increase of the spectrum at higher frequencies was probably a result of the energy input from the un-shielded compressor control unit. Since the major contributions are covered by the lower frequency range, the exact choice of the filter limiting frequency was not essential for estimating the turbulence level Tu. Setting limiting frequency values of 1 kHz or 2 kHz led essentially to identical results regarding turbulence intensity and other turbulent quantities. For this reason, the 2 kHz lowpass filter was chosen for the investigations presented here.

The great potential of HWA for the investigation of turbulent compressible non-ideal gas flows was checked by obtaining the autocorrelation $f(r/\lambda)$ for air and Novec™ 649 at several Mach numbers (assuming the Taylor hypothesis). The autocorrelation f enables a determination of the turbulent micro length scale λ and the integral length scale Λ (Rotta 1972). The results for the present test section are shown in Fig. 7. The experimentally obtained autocorrelation agreed well with the theoretical models proposed by Loisianskii and Birkhoff for (incompressible) isotropic turbulence. This might be interpreted as a strong indication that the well-established turbulence models

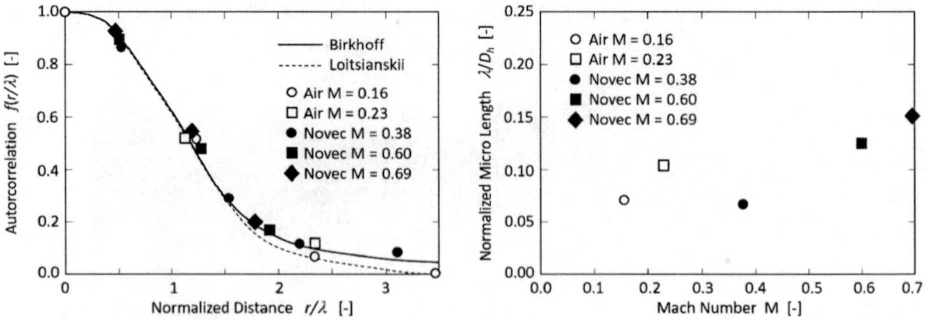

Fig. 7. Autocorrelation for Novec™ 649 and air at different Mach numbers (left) and turbulent micro length scale λ (normalized by means of the hydraulic diameter of the test section) against Mach number (right).

employed in CFD are also applicable for modelling high subsonic turbulent non-ideal gas flows. The turbulent micro length scale λ increased with Mach number M. Its value of $\lambda/D_h \approx 0.07$ (with D_h as the test sections hydraulic diameter) at very low Mach number, agreed well with observations for incompressible pipe and duct flows.

4 Conclusion and Outlook

The above experimental investigations demonstrated that hot-wire probes can withstand the considerable dynamic loads in high subsonic flows of heavy organic vapors at elevated density and temperature levels. This observation is generally promising for future experimental turbulence research dealing with non-ideal gas flows. Furthermore, the frequently employed laminar heat transfer correlation (1) proofed its reliability for high subsonic organic vapor flows, and static calibration of CTA in organic flows was conducted. The performance of the CTA system was assessed by obtaining turbulence levels in the wind tunnel and comparing the results with the outcome of an independent turbulence sphere test.

However, the hot-wire study presented here is far away from being completed and the authors recognized that hot-wire measurements can become extremely challenging in compressible organic vapor flow. Several relevant questions regarding signal processing and data reduction are still open. Furthermore, the design of hot-wire probes might be improved in order to provide higher spatial and temporal resolution.

The authors would be delighted if this first study could invite HWA experienced readers and experts to work on this interesting field. CLOWT has been intended to provide a test facility for collaborations, and the authors would highly appreciate such a collaboration with other researchers.

Acknowledgments. The support by SVMtec and the fruitful discussions about HWA with Dr. Kaiser, SVMtec, are highly acknowledged.

The authors would like to express their special thanks to the (unknown) reviewer who provided highly valuable suggestions for future work and whose review comments were very helpful for improving the quality of the paper.

References

Bruun, H.: Hot Wire Anemometry: Principles and Signal Analysis. Oxford University Press, Oxford (1995)

Barlow, J.B., Rae, W.H., Pope, A.: Low Speed Wind Tunnel Testing, 3rd edn. Wiley, New York (1999)

Horstman, C.C., Rose, W.C.: Hot wire anemometry in transonic flow. AIAA J. **15**, 395–401 (1977)

Colonna, P., Casati, E., Trapp, C., Mathijssen, T., Larjola, J., Turunen-Saaresti, T., Uusitalo, A.: Organic rankine cycle power systems: from the concept to current technology, applications, and an outlook to the future. ASME J. Eng. Gas Turbines Power **137**(10), 100801 (2015)

Kovasznay, L.S.G.: The hot-wire anemometer in supersonic flow. J. Aero. Sci. **17**(9), 565–572 (1950)

Macchi, E., Astolfi, M.: Organic Rankine Cycle (ORC) Power Systems. Elsevier, Amsterdam (2017)

Morkovin, M.V.: Fluctuations and hot-wire anemometry in compressible flows. Report North Atlantic Treaty Organization, Advisory Group for Aeronautical Research and Development (1956)

Reinker, F., Kenig, E.Y., aus der Wiesche, S.: Closed loop organic vapor wind tunnel CLOWT: commissioning and operational experience. In: Proceeding ORC 2019 Athens, Greece, paper-ID 47 (2019)

Reinker, F., Kenig, E.Y., aus der Wiesche, S.: CLOWT: a multifunctional test facility for the investigation of organic vapor flows. In: Proceedings ASME 2018 5th Joint US-European Fluids Engineering Division Summer Meeting, Montreal, Canada (V002T14A004) (2018)

Rotta, J.C.: Turbulente Strömungen. Teubner, Stuttgart (1972)

Smolyakov, A.V., Tkachenko, V.M.: The Measurement of Turbulent Fluctuations. Springer, Berlin (1983)

De Souza, F., Tavoularis, S.: Hot-wire response in high-subsonic flow, conference paper AIAA-99–0310 (1999)

Performance of a Rotatable Cylinder Pitot Probe in High Subsonic Non-ideal Gas Flows

Felix Reinker(✉), Robert Wagner, Max Passmann, Leander Hake, and Stefan aus der Wiesche

Muenster University of Applied Sciences, 48565 Steinfurt, Germany
{f.reinker, r.wagner, max.passmann, lh105490, wiesche}@fh-muenster.de

Abstract. The performance of a rotatable Pitot cylinder probe was experimentally investigated in the calibration test section of a closed loop organic vapor wind tunnel with the working fluid Novec™ 649 at elevated density level. Subsonic up to high subsonic flow conditions ($M = 0.5 - 0.7$) were covered by the experiments. The data reduction process for obtaining Mach numbers and velocity based on non-ideal flow calculations and corresponding isentropic relations were reviewed. The theoretical results were compared with experimental data. It was found that in the case of high subsonic non-ideal gas flows, significant deviations in comparison with conventional perfect gas flows occurred.

Keywords: Experimental techniques · Pitot probe · Non-ideal gas flows · Cylinder pressure distribution · Mach number

1 Introduction and Scope of Research

The stagnation (or total) pressure is the pressure measured by an instrument that brings the flow isentropically to rest. Such a device is usually called a Pitot probe, and its history can be traced back into the year 1732. Several probe geometries and combinations with static pressure taps are now available for obtaining Mach number and velocities in compressible flows as discussed by John and Keith (2006). In the case of turbomachinery applications, cylinder probes are also common, see Wyler (1975).

The theory of Pitot probes and the corresponding data reduction process, e. g. the famous Rayleigh-Pitot formula, are well treated in textbooks in the case of a perfect gas. Then, analytical expressions can be derived which make the use of Pitot probes for flow investigations rather easy. However, much less is known about the performance of Pitot probes in non-ideal gas flows, and further research is necessary. The European Community has funded recently an ERC Proof of Concept project called PROVA (Pitot pRobe for Organic Vapours) which has been started at the CREA laboratory (crealab.polimi.it) of Politecnico di Milano. The objective of that multi-year project is to design, build, and calibrate a Pitot probe for flow velocity and Mach number measurements in subsonic and supersonic flows of organic vapors which might be relevant to organic Rankine cycle (ORC) power systems for renewable energy sources.

M. Pini et al. (Eds.): NICFD 2020, ERCOFTAC Series 28, pp. 144–152, 2021.
https://doi.org/10.1007/978-3-030-69306-0_15

As an independent contribution to the research on Pitot probe applications in organic vapor flows, the outcome of an experimental study conducted at Muenster University of Applied Sciences are presented in the following. The present work was intended to assess the performance of a rotatable cylinder Pitot probe in high subsonic flows at a Mach number range of $M = 0.5 - 0.7$, where probe calibration coefficients remain still nearly constant. As discussed in detail by Langford et al. (1982), substantial disturbances due to the orientation of the stem can occur at higher Mach number levels. Such additional difficulties can be excluded, if the measurements are restricted to high subsonic flow conditions. It is guessed, that this might be a good starting point for future research including higher Mach number levels as well.

The present study was part of a preliminary investigation in order to establish measurement techniques for the determination of Mach numbers in high-subsonic and transonic organic vapor flow fields, as they are relevant for turbines. The findings of this study will be utilized in a parallel running project where a linear blade cascade, manufactured by selective laser melting process, is investigated in the Closed-Loop Organic vapor Wind Tunnel (CLOWT) at up to transonic flow conditions. The goal of that project is the review of high-subsonic and transonic loss correlations, which are already established for steam and gas turbines but not for non-ideal gas flows. High supersonic inflow conditions are excluded in the following because these regimes are out of scope of the ongoing research at Muenster (and also out of the working area of CLOWT).

2 Gasdynamics of Pitot Probes

The isentropic flow relations are fundamental for the data reduction process using Pitot probe pressure, p_0, and static pressure p. In subsonic perfect gas flow, the freestream flow Mach number M can be directly determined from the measured total pressure p_0 (Pitot probe pressure) and free static pressure p using the well-known isentropic flow relation (John and Keith 2006).

$$M^2 = \frac{2}{\gamma - 1}\left(\left(\frac{p_0}{p}\right)^{(\gamma-1)/\gamma} - 1\right). \tag{1}$$

In the case of supersonic perfect gas flow, the Pitot probe acts as a blunt-nosed body, and there will be a detached bow shock in front of the probe. Then, it is necessary to include also the normal shock relations into the analysis, and the well-known Rayleigh-Pitot formula

$$\frac{p_0}{p} = \frac{\gamma + 1}{2} M^2 \left(\frac{(\gamma + 1)^2 M^2}{4\gamma M^2 - 2(\gamma - 1)}\right)^{1/(\gamma-1)} \tag{2}$$

results in the case of a perfect gas with isentropic exponent γ. In Eq. (1), p is the free-stream static pressure and M the free-stream Mach number upstream of the bow shock. The Pitot probe pressure p_0 is the total pressure downstream of the shock.

The data reduction process described above can be easily adapted to non-ideal gas flow calculations. In this case, it is necessary to derive isentropic flow relations resting on suitable equations of state or on database information (e.g. REFPROP). The isentropic flow relations can be found by solving the mass and energy balance equations for a given initial state under the constraint of an isentropic expansion (s = constant). A corresponding calculation method including also the determination of normal shock relations was proposed by Passmann et al. (2017). Significant deviations were observed between non-ideal gas models and an equivalent perfect gas model. This is illustrated by means of Fig. 1 where the Mollier diagram of Novec™ 649 (the working fluid of the present study) and the impact of the thermodynamic model on the isentropic relation (1) are shown for an assumed expansion from the initial state T_0 = 400 K and p_0 = 5 bar. The deviations between the two models are of order 4 up to 8% demonstrating the need to capture the non-ideal gas behavior even in the high subsonic flow regime.

3 Experimental Setup

The present experiments were performed in the high-subsonic flow calibration test section of the closed-loop organic vapor wind tunnel (CLOWT, see Fig. 2) at Muenster University of Applied Sciences with the perfluorinated ketone Novec™ 649 as working fluid. Details on the test facility CLOWT and its working principle can be found in previous publications (Reinker et al. 2018; Reinker et al. 2019).

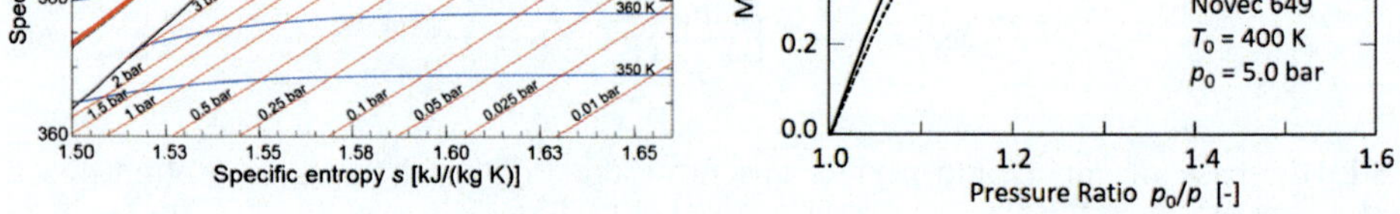

Fig. 1. Mollier diagram for NOVEC™ 649 (left) and subsonic isentropic flow relation for an expansion from T_0 = 400 K and p_0 = 5 bar obtained for a non-ideal gas model (REFPROP data) and equivalent perfect gas model (right).

After passing the compressor with running speed n, the organic vapor was decelerated in the diffuser and entered the settling chamber equipped with a chiller where total temperature T_0 (uncertainty $\Delta T/T = 0.17 - 0.19\%$) and stagnation pressure p_0 were measured. It should be noted that pressure measurements in organic vapor flows are generally challenging. It is not only the high temperature of the organic vapor which has to be decoupled from the pressure transducer, but condensation inside of the lines has to be avoided, too. A possible strategy to handle this issue is to purge the pressure lines and hence ensure that there is no fluid at the liquid state inside of the lines. However, this strategy requires substantial efforts, and active purging complicates the measurement procedure since an additional auxiliary system has to be controlled.

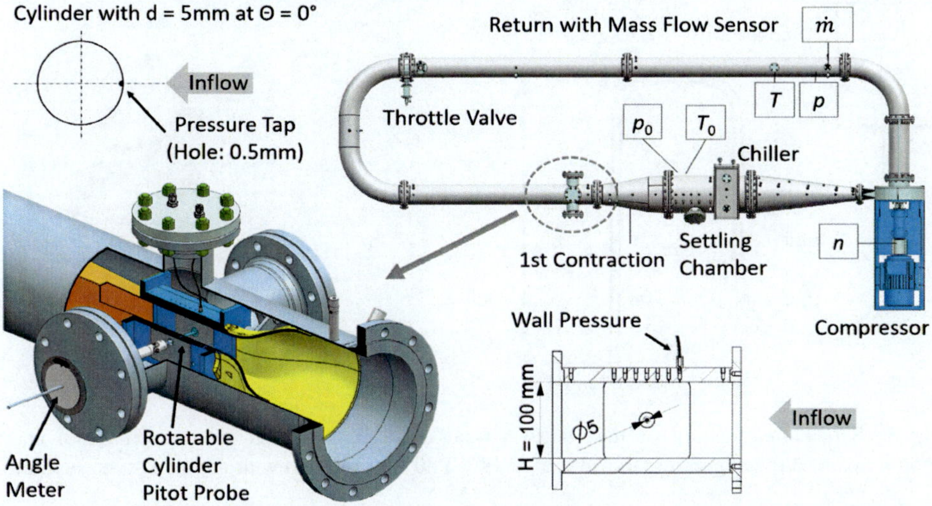

Fig. 2. Wind tunnel test facility CLOWT with instrumentation (top right), high subsonic calibration test section dimensions (lower right), look inside of the test section assembly (lower left), and dimensions and circumferential angle definition of the rotatable cylinder with pressure tap (top left).

For simplicity reasons and to avoid such issues, CLOWT was equipped with a passive pressure transducer thermal decoupling system (according to the German industrial standard DIN 16282), which is technically based on the well-known siphon-effect using a U-tube (see Fig. 3). During operation, the thermal decoupling of the pressure transducers worked excellent so that the transducers temperature was equal to the ambient temperature. But in turn, this rather simple approach might introduce systematic measurement errors. Figure 3 shows on the right a photograph of our first utilized measurement device including a pressure transducer (WIKA P30 0–6 bar absolute pressure) and the thermal decoupling siphon system. On the left side of Fig. 3, the corresponding schematic drawing illustrates the physical source of measurement error: Due to the low temperature level close to the transducer, local condensation of the vapor in the siphon could occur. In this case the geodetic height h creates an unwanted offset in the measured pressure p. This systematic error source can be reduced by the minimization of the siphon height h_{max}. Another, somewhat related, error source might occur due to misalignment between transducer diaphragm and the maximum turning point of the U-tube. But this issue can be eliminated by careful alignment of the components. As consequence, the first utilized thermal decoupling siphon system, which was not optimized in terms of the height h_{max}, might contain a systematic error of about 1500 Pa, resulting in an overall uncertainty of $\Delta p/p = 0.6 - 1.5\%$ for the measurements.

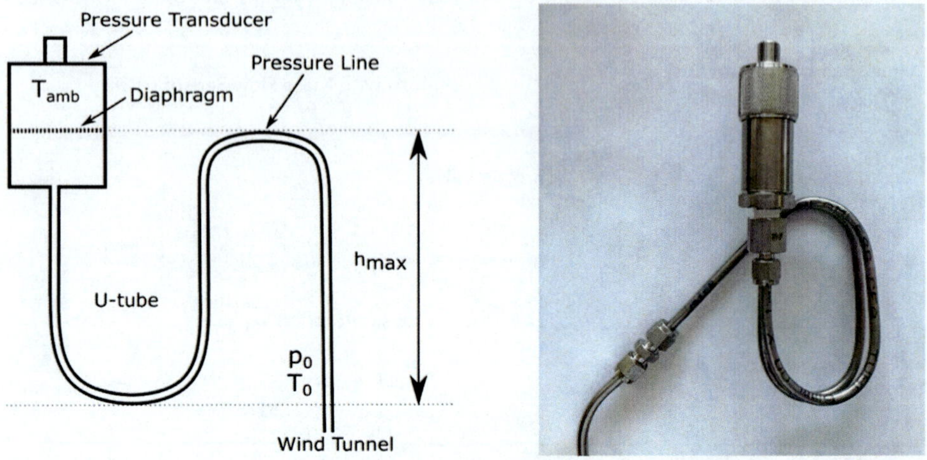

Fig. 3. Schematic drawing of the static thermal decoupling siphon system (left side) and photograph of utilized pressure transducer (WIKA P30 0–6 bar abs) with double U-tube syphon.

The mass flow rate $\dot{m}$ through the test section was recorded independently by means of a mass flow device with a level of uncertainty of $\Delta\dot{m}/\dot{m} = 1.6 - 4.5\%$ in the subsonic return of the closed-loop wind tunnel, where also pressure and temperature were recorded for control purposes.

In the calibration test section, the free static pressure p was measured by means of a wall pressure tap upstream of the test object (see Fig. 2 at the lower right sketch). CLOWT is designed with a two-stage contraction. The first subsonic axisymmetric nozzle (standard diameter DN500 to DN250) offers a moderate contraction ratio of about 3.7, whereas the second contraction accelerates the flow to high subsonic or transonic flow depending on the chosen individual nozzle device. In the present study, the second contraction was achieved by a three-dimensional nozzle based on additive manufacturing providing the cross-section change from round to rectangular contraction (DN250 to 50 mm $\times$ 100 mm) leading to a total contraction ratio of about 38. This nozzle was designed to achieve high subsonic flows up to $M = 0.7$. The test section inflow turbulence level was measured by means of hot-wire anemometry, and it was found that it was of order 0.5 up to 1.0% (for $M = 0.16$ up to 0.7) with a turbulent micro length λ of about 10% in respect to the hydraulic diameter of the test section.

A large-scale model of a cylinder Pitot probe with diameter $d = 5$ mm was mounted in the calibration test section (see Fig. 2 left side). The probe blockage was $d/H = 5\%$ in the calibration test section. The cylinder probe was equipped with a small pressure hole (0.5 mm) for measuring the cylinder wall pressure p_w (equal to p_0 at the stagnation point at $\theta = 0°$), and the probe could be precisely rotated from the outside of the wind tunnel by means of a rotatable sealed shaft and an angle meter. This enabled recording of the entire circumferential pressure coefficient distribution C_p and an assessment of the probe sensitivity regarding flow alignment. In the present study the Pitot configuration with pressure tap at the front stagnation point is defined as the circumferential angle $\theta = 0°$.

The following experiments were conducted at nearly constant total temperature $T_0 = 370 \pm 1$ K and constant compressibility factor $Z = 0.88 \pm 0.01$ in the test section. Due to fundamental thermodynamics, the organic vapor states vary through the operational conditions in the wind tunnel. Hence, the total pressure p_0 and the total density varied during the experiments in order to ensure a constant value for Z (the actual density range was between $\rho_0 = 32$ and 38 kg/m^3). The actual Mach number M was controlled by the compressor running speed n (at full throttle valve). For each Mach number M, a corresponding Reynolds number Re resulted. The Reynolds number range was between $Re = 1.8 \times 10^5$ up to 6.1×10^5 corresponding to transitional values for cylinders depending on the actual Mach number level.

In turbomachinery applications, the probes are typically large compared to the flow area, and the question regarding suitable blockage corrections arise in order to obtain the correct flow parameters based on the measured, uncorrected values (denoted by the subscript *Pitot* in the following). This was also the case in the present experimental study (with an area blockage factor $d/H = 0.05$). Based on Wyler (1975), a velocity blockage factor

$$\varepsilon = \frac{w - w_{Pitot}}{w_{Pitot}} = \frac{C_D}{2(1-M^2)}\frac{d}{H} \quad \text{and} \quad C_p - 1 = \left(\frac{w_{Pitot}}{w}\right)^2 \left(C_{p,Pitot} - 1\right) \tag{3}$$

for the subsonic compressible flow was considered for data reduction for obtaining the "true" velocity w and pressure coefficient C_p. The validity of ε for non-ideal gas flows will be proofed in a separate study. It has to be mentioned that in dense gas flows, the probe Reynolds number Re (based on probe diameter d) is typically rather high, and a correct value for the drag coefficient C_D has to be used. In conventional applications, it is typically assumed that C_D corresponds to the well-known subcritical expression (Wyler 1975; Morrison et al. 1963), but this was not the case in the present investigations, and supercritical or transitional values were used (see results of the next section).

4 Results and Discussion

Firstly, the robustness of the cylinder Pitot probe against flow misalignment (i.e. incidence angle $\Delta\theta$) were tested. The results are shown in Fig. 4 (left). At low incidence, the agreement between the stagnation pressure p_0 measured at the settling chamber and the Pitot probe pressure p_w were well within the uncertainty level of the employed pressure transducers which were of order 0.2%. Systematic deviations occurred for larger flow misalignment, i.e. incidence. That was expected, and it is rather typical for cylinder probes. However, a slight asymmetry was observed regarding positive and negative incidence at higher Mach number levels. That asymmetry was explained by imperfections of the actual cylinder geometry and non-perfect edges of the pressure hole. This assumption could be confirmed by a microscopic image at 200x magnification (see Fig. 5). However, the cylinder probe worked well within a $\pm$ 10° range.

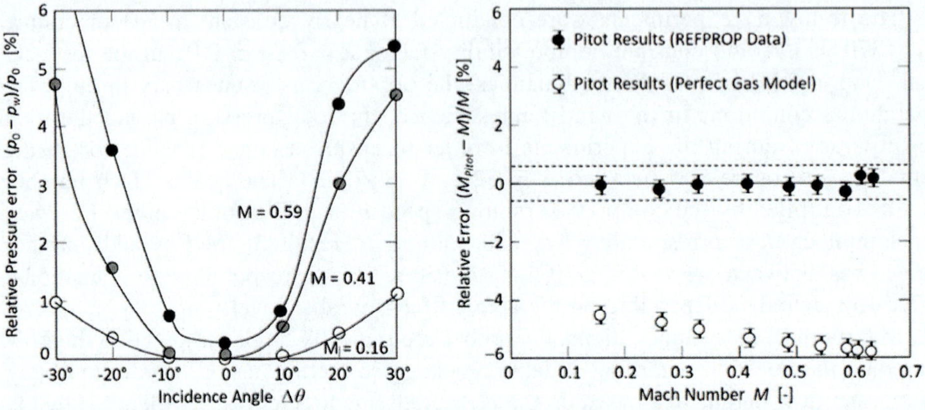

Fig. 4. Experimental results for the cylinder Pitot probe. Robustness against flow misalignment (left) and accuracy of Mach number calculations based on Pitot probe data and thermodynamic models (right).

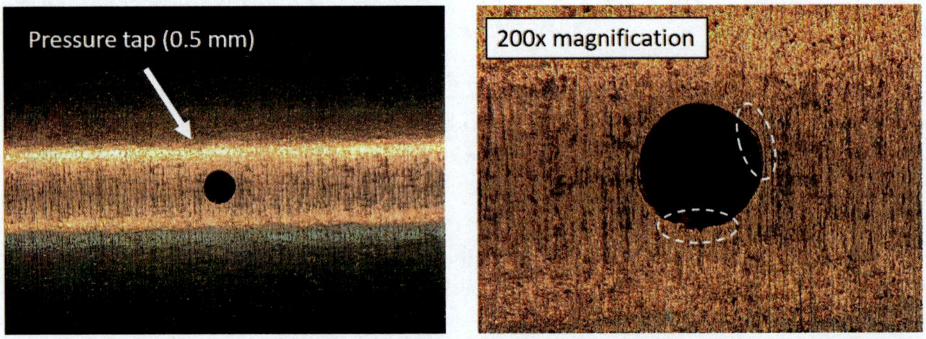

Fig. 5. Illustration of the imperfection of the cylinder pressure tap at 200× magnification, explaining the slight asymmetry in the pressure distribution of upper and lower side of the cylinder.

Secondly, the Pitot pressure signal at stagnation point (i. e. at $\theta = 0°$ yielding $p_w = p_0$) was recorded for several Mach numbers M and a constant compressibility factor $Z = 0.88$. By means of recorded Pitot pressures and static test section wall pressures, Mach numbers M_{Pitot} were calculated using REFPROP data and using an equivalent perfect gas model, Eq. (1). The Pitot-Mach numbers M_{Pitot} were compared with the reference Mach number M obtained independently using the mass flow rate sensor and settling chamber data providing the reference total pressure and temperature values. The total uncertainty level of the reference Mach number M was of order 0.6%. An excellent agreement was found between the reference Mach numbers and the Pitot-based Mach numbers obtained by means of non-ideal gas data (REFPROP). The deviations were well within the experimental uncertainty level, see dotted lines in Fig. 4 (right). As expected, see Fig. 4, the Mach number predictions based on an equivalent perfect gas model were somewhat poorer. Even in the moderate subsonic flow regime, the deviations were found to be significant. This observation supports the findings illustrated in Fig. 1. In the case of organic vapor flows, it is essential to employ appropriate equation of states for data reduction even in the subsonic flow case.

Finally, the pressure coefficient distributions $C_p = (p_w(\theta) - p)/(p_0 - p)$ against the full circumferential angle θ were obtained for some Mach and Reynolds numbers for which, depending on the Mach number, the transition to turbulence was expected, see Fig. 6 (left). The results for $M = 0.17$ agreed well within literature data obtained for essentially incompressible flow. For $M = 0.60$ locally supersonic flow was achieved for $C_p < -1.5$, and a significant drag divergence was observed for that super-critical Mach number (in terms of transition). The circumferential angle θ_{cr}, for which $C_p = 0$ occurred, depended on the Mach number as firstly observed by Wyler, 1972. The first experimental data seem to indicate that this quantity could be scaled by means of a Prandtl-Glauert rule, Fig. 6 (right). However, at the present moment no definite answer can be given to the interesting question whether this observation might be used for measuring static pressures as well. A corresponding study about the theoretical principles of cylindrical pneumatic probes by Birri and Voegtli (1987), discussed in detail the problems.

Although probe stem diameters of 5 mm are not uncommon in practice, it is remarkable that this dimension leads already to supercritical flow past a cylinder and drag divergence in the case of high subsonic organic vapor flows. This important fact has to be considered for calculating accurate blockage corrections and data reduction in non-ideal gas flow experiments.

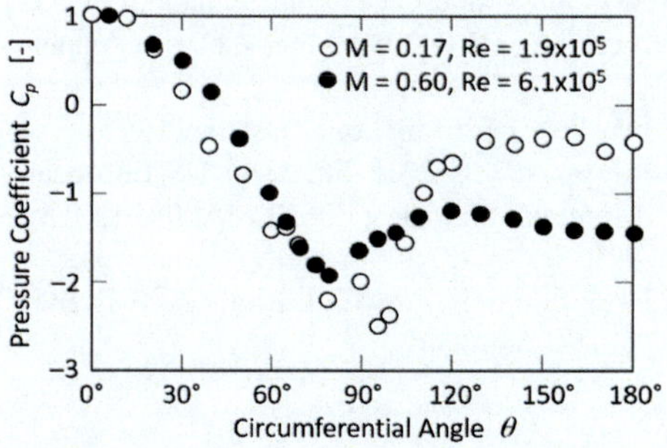

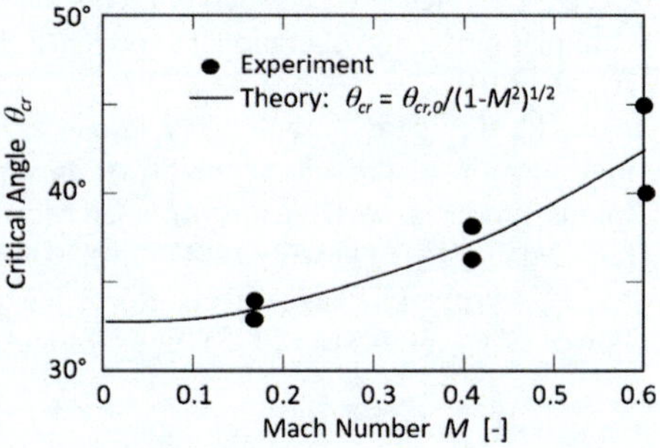

Fig. 6. Cylinder pressure distribution in Novec™ 649 at Z = 0.88 (left) and critical circumferential angle (right).

5 Summary

The performance of a rotatable Pitot cylinder probe was experimentally investigated in the calibration test section of a closed loop organic vapor wind tunnel with the working fluid Novec™ 649 at elevated density level. Subsonic up to high subsonic flow conditions (M = 0.5 – 0.7) were covered by the experiments. The data reduction process for obtaining Mach numbers and velocity based on non-ideal flow calculations and corresponding isentropic relations were reviewed.

It was found that in the case of high subsonic non-ideal gas flows, significant deviations in comparison with conventional perfect gas flows occurred, and that data reduction processes using simplified equivalent ideal gas models lead to significant deviations. Furthermore, it was noticed that even in the case of typical, small stem diameters, supercritical flow conditions occur in dense gas flows. This has been considered for obtaining accurate blockage corrections for probes in test sections.

References

Birri, I., Voegtli, D.: Theoretical principles of the cylindrical, 2-hole pneumatic probe. In: Proceedings ASME Turbo Expo (paper-ID 87-GT-151) (1987). https://doi.org/10.1115/87-GT-151

John, J.E., Keith, T.G.: Gas Dynamics. Pearson Prentice Hall, Upper Saddle River (2006)

Langford, R.W., Keeley, K.R., Wood, N.B.: Investigation of the transonic calibration characteristics of turbine static pressure probes. In: Proceedings ASME Turbo Expo (V001T01A093) (1982). https://doi.org/10.1115/82-GT-280

Morrison, R.B., Ingle, M.J., Radbill, J.R.: Design data for aeronautics and astronautics. J. Appl. Mech. **30**(1), 159 (1963). https://doi.org/10.1115/1.3630094

Passmann, M., aus der Wiesche, S., Joos, F.: A one-dimensional analytical calculation method for obtaining normal shock losses in supersonic real gas flows. In: IOP Conference Series: Journal of Physics: Conference Series, vol. 821, p. 012004 (2017). https://doi.org/10.1088/1742-6596/821/1/012004

Reinker, F., Kenig, E.Y., aus der Wiesche, S.: Closed loop organic vapor wind tunnel CLOWT: commissioning and operational experience. In: Proceeding ORC 2019 Athens, Greece, paper-ID 47 (2019)

Reinker, F., Kenig, E.Y., aus der Wiesche, S.: CLOWT: a multifunctional test facility for the investigation of organic vapor flows. In: Proceedings ASME 2018 5th Joint US-European Fluids Engineering Division Summer Meeting, Montreal, Canada (V002T14A004) (2018). https://doi.org/10.1115/FEDSM2018-83076

Wyler, J.S.: Probe blockage effects in free jets and closed tunnels. ASME J. Eng. Gas Turbines Power **97**(4), 509–514 (1975). https://doi.org/10.1115/1.3446046

The Effect of Isentropic Exponent on Supersonic Turbine Wakes

David Baumgärtner, John J. Otter(✉), and Andrew P. S. Wheeler

University of Cambridge, Cambridge, UK
{db701,jjo26,aw329}@cam.ac.uk

Abstract. The isentropic exponent is known to have a large impact on turbine aerodynamics yet there is little research into how this impacts the development of turbine wakes. This study aims to explore how the aerodynamics of a supersonic turbine cascade are affected by the isentropic exponent. This is achieved through both experimental and computational methods with working fluids of air, CO_2 and R134a which gives a range of isentropic exponents from 1.07 to 1.4. The experiments are performed using a newly modified transient wind tunnel with the capability to obtain total pressure measurements in the wake of the cascade. This paper describes the development of the experimental apparatus as well as reporting the first-of-their-kind wake measurements from a turbine cascade operating with R134a. The resultant experimental wake measurements serve as a validation data set for various computational fluid dynamic approaches.

Keywords: Turbine cascade · Experimental · ORC · R134a

1 Introduction

The isentropic exponent is a key factor in turbine design as it governs the inviscid gas dynamics for a particular working fluid. Within this work, the isentropic exponent is defined based on the variation of pressure and density across an isentropic expansion (Eq. 1).

$$\left(\frac{\partial p}{\partial \rho}\right)_{\mathrm{s}} = k\frac{p}{\rho} \tag{1}$$

The interest in the application of turbomachinery with working fluids other than air or steam is increasing due to the wider industrial application of heat recovery cycles such as Organic Rankine Cycles. Such cycles feature working fluids with low isentropic exponents, often with values less than 1, and the aerodynamic behaviour of turbomachines operating with such unconventional working fluids is not well understood.

Experimental studies into the behavior of high-molecular weight fluids operating under non-ideal conditions has been undertaken for supersonic converging-diverging nozzle geometries [2–4]. Such experimental data is essential to validate thermodynamic models and computational fluid dynamic models. However, while nozzle flows can capture the bulk expansion through a heat recovery

M. Pini et al. (Eds.): NICFD 2020, ERCOFTAC Series 28, pp. 153–161, 2021.
https://doi.org/10.1007/978-3-030-69306-0_16

turbine nozzle, they do not provide insight into the detailed aerodynamics of turbomachinery flows. For a turbine nozzle, the most notable feature that cannot be determined from nozzle testing is the wake, and hence loss, downstream of a turbine vane. Therefore, the aim of this paper is to explore how isentropic exponent affects turbine wake flows. This is achieved through a combination of experimental and computational approaches.

2 Methodology

2.1 Experimental Setup

The experiments reported within this paper were conducted with a transient wind tunnel, otherwise known as a Ludwieg tube. The major advantage of operating with such an arrangement is that the tunnel working fluid can be changed without any modifications to the tunnel. Details of the full set up of this tunnel have been reported previously [1] and hence for brevity this paper only reports the changes made to this rig to obtain total pressure measurements.

The working section consisted of an annular cascade with a design Mach number of 1.45, exit flow angle of 75° and isentropic exponent of 1.4. The vane was designed using the design method developed by Wheeler et al. [5] and is geometrically similar to a stator vane found in a typical Organic Rankine Cycle (ORC) stage [6]. The cascade was formed of eight blades with a hub diameter of 30 mm, a shroud 40 mm and hence a span of 5 mm. The cascade is mounted between an elliptic hub and a tapered afterbody (Fig. 2).

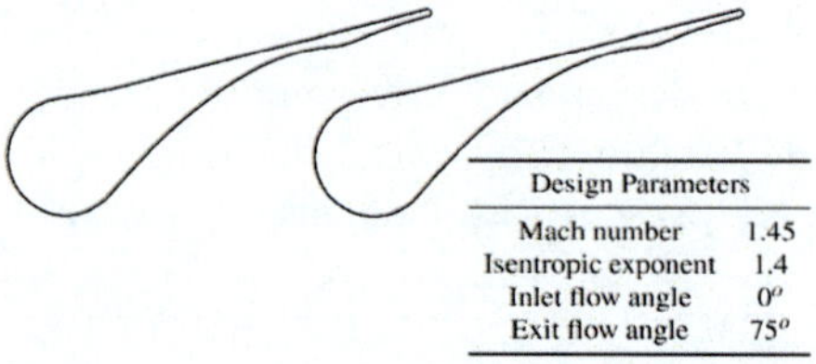

Fig. 1. Baseline vane design

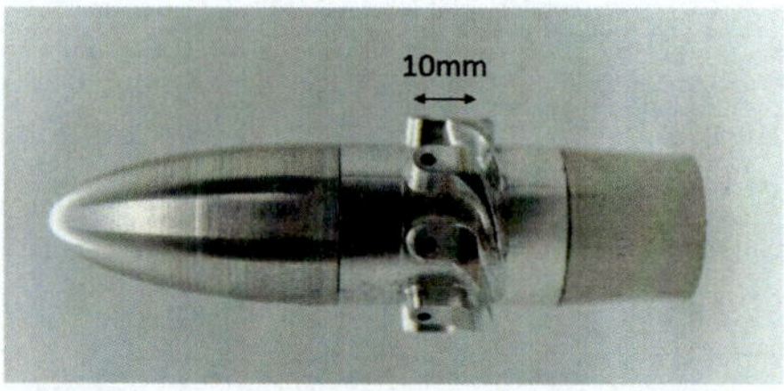

Fig. 2. Cascade geometry mounted between elliptic hub and afterbody

The expansion of the working fluid throughout the vane was measured by placing eight pressure tappings along the centre line of the vane. Each pressure tapping has a diameter of 0.5 mm. The first pressure tapping is located just upstream of the throat, and the last pressure tapping is located at 12% of the axial chord downstream of the trailing edge. Under the assumption of identical flow through each of the passages, the actual measurements were spread out over multiple passages due to space limitations. In order to increase the density of the experimental data points on the shroud of the cascade, the test section was

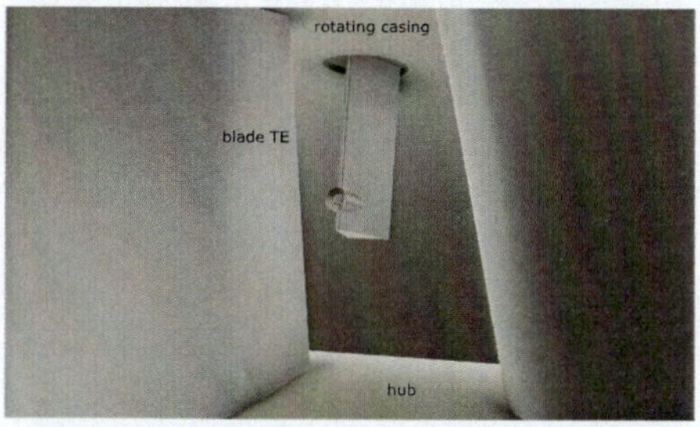

Fig. 3. Schematic view of the wedge probe

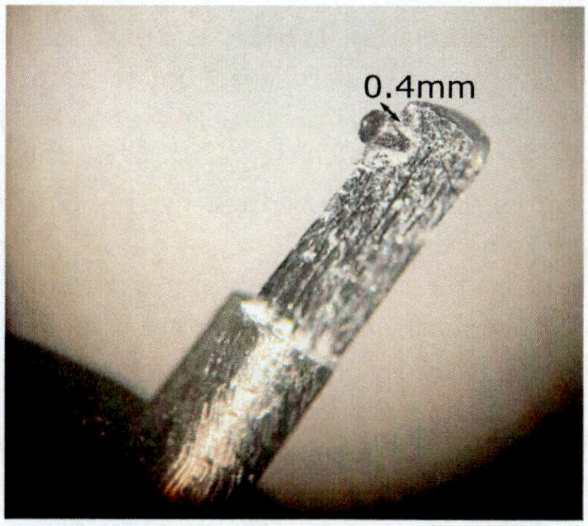

Fig. 4. Picture of the wedge probe

designed in such a way that the outer casing with the pressure tappings could be rotated circumferentially relative to the cascade. This rotation of the casing can be done in increments of 2°.

Besides the wall static pressure measurements, an additional probe was designed to measure the wake profile at mid-span in the supersonic region downstream of the vane. In a first iteration of the design, a circular probe was investigated. While this design was easy to manufacture, it came with the cost of a high degree of interaction with the flow in the free stream. The final probe consisted of a two parts: a wedge section which was machined from a single piece of aluminium and a hypodermic tube which was which was glued into the wedge. This hypodermic tube was aligned so as to be parallel to the flow and had an outside diameter 0.4 mm and inside diameter of 0.2 mm (see Figs. 3 and 4). For reference, the trailing edge thickness of the vane was 0.25 mm.

Making use of the rotational casing, the probe could be used to obtain a wake profile. In order to reduce the number of consecutive runs of the experiment, the wake profile was only recorded at one angular position of the probe, the free stream velocity angle. This velocity angle was obtained by positioning the wedge in the free stream of the vane and to rotate the probe around its own axis to find the angle of maximum pressure. CFD simulations have shown that the variation of the velocity angle along the circumferential measurement positions is within ± 5°.

A total of three different working fluids were tested: air, CO_2 and R134a with a summary of the inlet conditions given in Table 1. These fluids were chosen as they present a range of isentropic exponents from 1.4 to 1.07. In addition, the fundamental derivative Γ and compressibility factor Z based on turbine inlet conditions are also reported. The total pressure p_0 and temperature T_0 at the inlet of the working section were varied for each working fluid for a number of practical reasons. For R134a a total pressure of 5 bar was chosen as this corresponds to the fluid saturation pressure at room temperature. Operation at this condition meant that a pump was not needed and the pressurised gas bottle could be directly connected to the tunnel. For both air and CO_2 the total pressure was chosen such that the operating Reynolds number Re was close to that of the R134a, which in this case was 1.6×10^6 based on the conditions at the vane exit and the axial chord. Prior to running with CO_2, the tunnel was heated 400 K so as to avoid any two-phase flow through the vane expansion.

Before charging the tunnel with R134a the entire tunnel was flushed with nitrogen and then vacuumed to a pressure of 2 Torr. Upon completion of a run a Yellow Jacket XLT2-AP recovery unit was used to the recover the R134a from within the tunnel. As a result, obtaining each measurement for R134a took around 4 hours, compared to air for which a run could be completed in 15 min.

Table 1. Experimental run conditions

	p_0 [bar]	T_0 [K]	Re	k	Γ	Z	PR
Air	8.2	293	1.6×10^6	1.40	1.21	1.0	4.75
CO_2	8.2	400	1.4×10^6	1.27	1.11	0.98	4.75
R134a	5.0	293	1.6×10^6	1.07	1.02	0.88	4.75

For this set of experiments, the cascade geometry remained fixed and the pressure ratio PR across the blades was held constant. The pressure ratio is defined as the ratio between inlet stagnation pressure and the static pressure that is the furthest downstream of the cascade. As a result of a constant pressure ratio and variable isentropic exponent, the exit Mach number for each blade was different for each fluid.

For the wake traverse with air as a working fluid, multiple measurements were taken at selected pitchwise traverse locations. An estimate for the experimental uncertainty has been calculated based on the variance of the normalised shocked total pressure for the pitchwise locations where there were multiple measurements (Eq. 2). Within Eq. 2 p_0^{shocked} is the value of normalised total pressure, $\overline{p_0^{\text{shocked}}}$ is the mean value of normalised total pressure at the given pitchwise location and n is the number of wake measurements. With air as a working fluid σ_{wake} was found to be 0.025 based on 48 measurements.

$$\sigma_{\text{wake}} = \sqrt{\frac{\sum_{i=1}^{n}(p_0^{\text{shocked}} - \overline{p_0^{\text{shocked}}})^2}{n-1}} \tag{2}$$

2.2 Computational Method

The experimental aspect of this work is supported with both 2D and 3D Reynolds-averaged Navier-Stokes Computational Fluid Dynamic simulations. The 3D calculations were undertaken based on the experimental vane geometry and with an aim to compare RANS solutions to experimental data. The results based on the experimental study cannot be used to completely isolate changes in isentropic exponent as the vane exit Mach number varied between each case. Therefore a 2D study was undertaken to explore how variations in the isentropic exponent at a constant vane exit Mach number affect the wake.

Within this study, the implicit density based solver from the commercial package ANSYS FLUENT V17.0 was used. Closure of the RANS equations was achieved with the Spalat-Allmaras model. Each working fluid was modelled using the inbuilt interface with the REFPROP Helmholtz free-energy equations of state. The computational meshes were generated with ANSYS Turbogrid and the 3D mesh of the experimental vane consisted of 2 million elements with 70,000 elements in the blade-to-blade plane. The 2D meshes consisted of 100,000 elements in the blade-to-blade plane. The wall distance for each mesh was set to give a y^+ of 30 such that near wall modelling was achieved with wall functions. Further details of this computational approach, including validation and verification, are reported by Baumgärtner et al. [1].

3 Results

3.1 Experimental Measurements of Wakes

As the total pressure probe is measuring within the supersonic region of the flow, a bow shock will form upstream of it. Therefore, the probe will measure stagnation pressure that is subject to some shock loss. As the Mach number upstream of the probe is not measured, correcting the measured stagnation pressure becomes challenging. As a result, it is not possible to make a direct comparison between the three fluids, as the shock loss will be different for all three fluids.

The location of the traverse relative to the vane is shown on the contour plot of Mach number for the example of air (top plot of Fig. 5). From the CFD solutions, it is possible to extract the Mach number at the position of the probe. With this pre-shock Mach number and the REFPROP equations of state, theoretical post-shock flow properties can be calculated based on the assumption of a normal shock (middle plot of Fig. 5). This is an approximation, as the probe is not located in a uniform flow field but rather in a complex 3D flow field where a bow shock is formed.

An additional challenge of the wake traverse was the small scale of the cascade and the concurrent small scale of the flow features such as the wake. In order to highlight the changes in the flow properties in the close proximity to the centre of the face of the probe, the minimum and maximum values around this point were extracted from the CFD. This was done by extracting the min/max values on a sphere around this point with a radius of 0.2 mm, equal to the inside diameter of the hypodermic tube which is parallel to the flow (Fig. 4). The extent of the variation to the shocked total pressure is shown as the shaded area in the middle plot of Fig. 5.

The wake profiles for the three fluids, with the measured stagnation pressure, are shown in the bottom plot of Fig. 5. Note that where multiple measurements were taken the mean value is presented. For all three fluids, the probe is able to

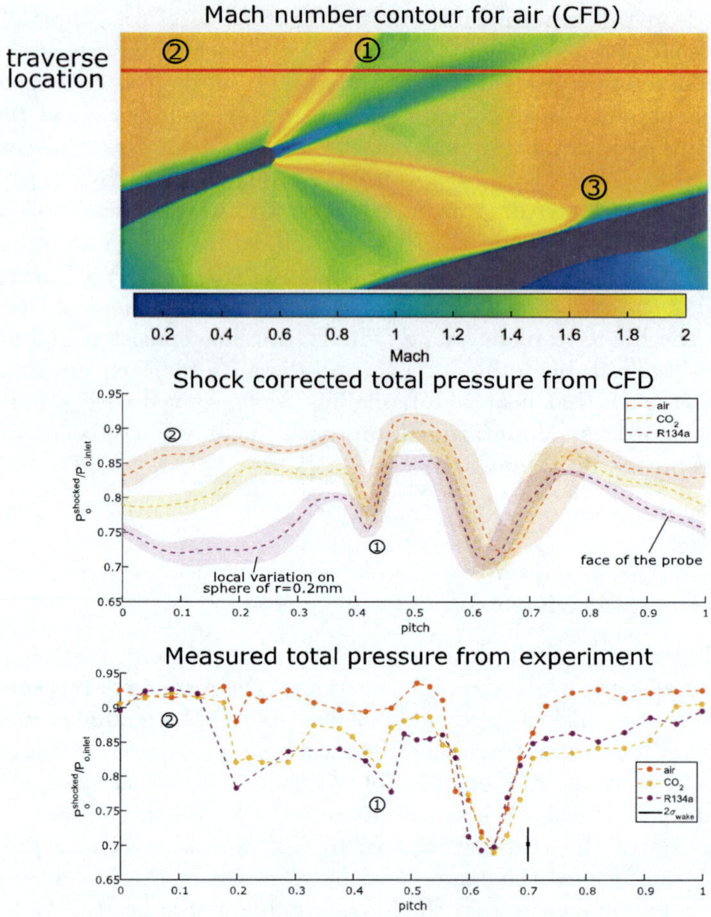

Fig. 5. Comparison of wakes extracted from CFD to experimental measurements

pick up the wake downstream of the cascade with about ten measurements (1° increments) inside the wake. At the centre of the wake, the measured shocked total pressure reads about 0.69 $p_{0,\text{in}}$. Looking at the free stream, air shows the highest measured total pressure between the three measured fluids. There are two possible reasons for this. Either a different Mach number in the free stream leads to higher shock losses and/or higher shock loss due to k at the same Mach number. The detailed wall static measurements for the different fluids did reveal that the unguided expansion for R134a (low k) is higher, and one can observe higher Mach numbers. Hence, one can expect the Mach number at the mid-span to behave similarly.

Looking at the shock-corrected computational results at the location of the wake traverse (bottom plot of Fig. 5), there is a good agreement with the experimental data in the region of the wake, specifically the trend of the wake depth and the width. Also, the dip to the left side of the wake (position (1)) shows the same trend in the CFD solution as the experimental findings. This dip is the result of the expansion fan and compression shock being reflected off the suction side of the trailing edge (top plot of Fig. 5). At this point, the probe sees a higher Mach number and thus a stronger bow shock is formed. The observed dip in measured pressure is increasing for lower values of k due to more substantial shock losses.

At the same time, there are also differences between the results of the CFD and the measured data. This is the case for the free stream (position (2)). One likely reason for this discrepancy between CFD and experimental data is that the CFD is not able to correctly predict the shock boundary layer interaction at position (3), which then sets the size of the separation bubble, which in turn affects the Mach number in the free stream.

3.2 Effect of Isentropic Exponent on Vane Wakes

In order to explore how isentropic exponent effects the vane wakes, a series of 2D simulations were undertaken. Within these simulations, the vane was redesigned for each working fluid, such that an exit Mach number of 1.5 was achieved. For each fluid, the inlet total conditions were varied, such that the Reynolds number, based on exit conditions and the vane axial chord, was constant between each of them with a value of 1.7×10^6. Therefore, changes in the flow physics can be attributed to the isentropic exponent. Figures 6 and 7 show the variation of Mach number and loss coefficient respectively at two axial cuts downstream of the trailing edge. The loss coefficient is defined as the difference between the enthalpy h and an isentropic enthalpy h_s, $\zeta = (h - h_s)/(h_0 - h_s)$. The local isentropic enthalpy was calculated using the relevant equation of state, local pressure and inlet entropy. The symbol h_0 denotes in the inlet stagnation enthalpy. Moreover, Fig. 8 shows distributions of Mach number and static pressure coefficient with the location of the aforementioned axial cuts marked on the top left sub-figure.

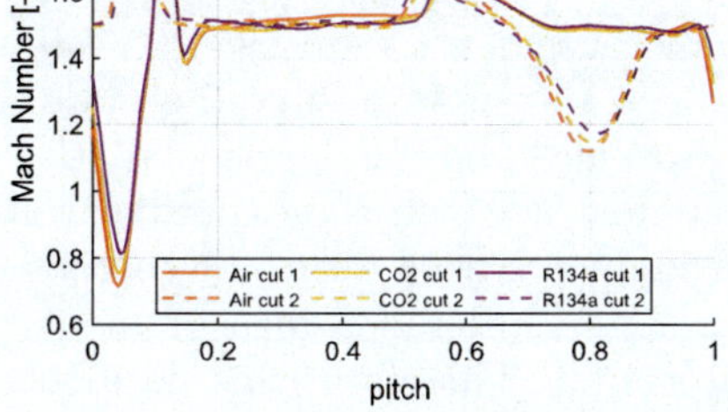

Fig. 6. Variation of mach number

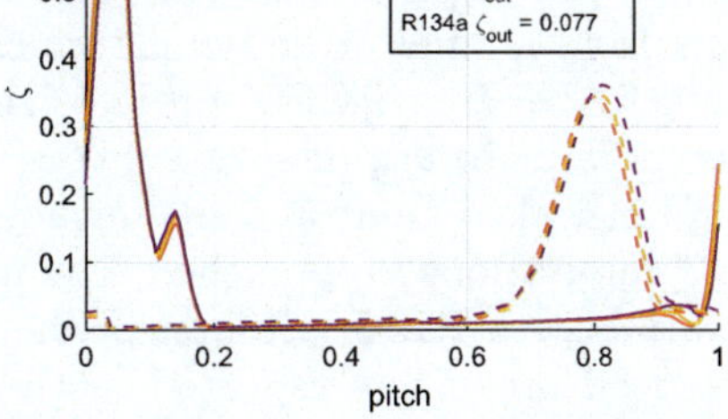

Fig. 7. Variation of local Loss Coefficient

At the cut closest to the trailing edge, the R134a wake is both narrower and shallower compared to the air and CO_2 wakes (solid lines in Fig. 6). This is to be expected as, due to the lower value of k, the size of the separated base region for R134a will be smaller than that of air and CO_2 [1]. In terms of loss coefficient, the distribution just downstream of the trailing edge is very similar between the fluids (solid lines in Fig. 7). At the location further downstream, the R134a wake still remains the shallowest as shown by dashed lines in Fig. 6. Although the wake is shallower for R134a, in terms of loss coefficient it displays the highest value (dashed lines in Fig. 7). When a mixed-out loss coefficient is evaluated, the loss forms a monotonic function of isentropic exponent with values of 0.0629, 0.0682 and 0.0767 for air, CO_2 and R134a respectively. From Fig. 8 it is clear that, compared to the air wake, the R134a wake is subject to a larger adverse pressure gradient between the trailing edge and downstream flow. It is postulated that the increased loss coefficient at cut 2 for the R134a wake is a result of this larger adverse pressure gradient.

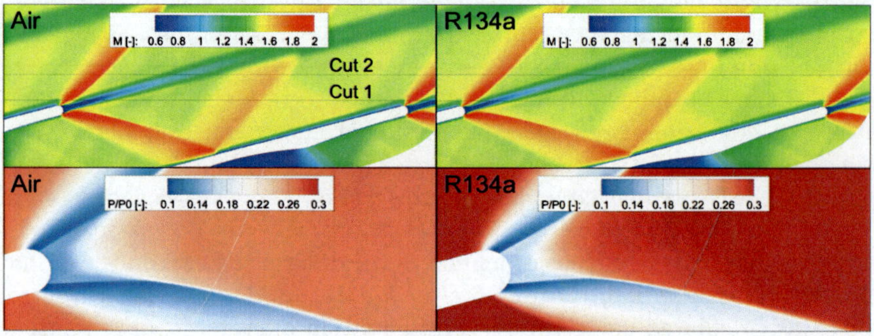

Fig. 8. Distributions of mach number and pressure coefficient (p/p_0) from 2D CFD

4 Conclusions

This paper has presented a methodology for obtaining detailed wake measurements from a small-scale wind tunnel. These measurements were used to validate a RANS based CFD method which has been shown to resolve the major differences in total pressure between various working fluids. In particular, the trends in wake extent and magnitude were matched between the experimental and computational approaches. The wake measurements from a turbine cascade operating with R134a are first-of-their-kind, as no such detailed measurements have been previously reported in the open literature. Therefore, this data set has the potential to be used as a key validation data set for other CFD methods. In order to facilitate this, the authors would like to share the experimental data and geometry so that it may be used for further validation purposes. Those interested should contact the authors to obtain the necessary data.

From the supporting CFD study, it has been shown that the wake immediately downstream of the vane trailing edge becomes shallower and narrower as the isentropic exponent is reduced. As the wakes mixes out immediately downstream of the trailing edge, lower isentropic exponents lead to a higher loss owing to a larger adverse pressure gradient downstream of the wake.

Acknowledgments. This work was supported by the EPSRC (EP/L027437/1) and computational resources were provided by EPSRC Tier-2 capital grant EP/P020259/1. The authors would like to thank the reviewers for their insightful comments and valuable feedback.

References

1. Baumgärtner, D., Otter, J.J., Wheeler, A.P.S.: The effect of isentropic exponent on transonic turbine performance. J. Turbomach. **142**(8) (2020). 081007
2. Reinker, F., Hasselmann, K., aus der Wiesche, S., Kenig, E.Y.: Thermodynamics and fluid mechanics of a closed blade cascade WindTunnel for organic vapors. J. Eng. Gas Turbines Power **138**(5) (2016). 052601
3. Robertson, M., Newton, P., Chen, T., Costall, A., Martinez-Botas, R.: Experimental and numerical study of supersonic non-ideal flows for organic rankine cycle applications. J. Eng. Gas Turbines Power **142**(8) (2020). 081007
4. Spinelli, A., Cammi, G., Gallarini, S., Zocca, M., Cozzi, F., Gaetani, P., Dossena, V., Guardone, A.: Experimental evidence of non-ideal compressible effects in expanding flow of a high molecular complexity vapor. Exp. Fluids **59**(8), 126 (2018)
5. Wheeler, A., Ong, J.: The role of dense gas dynamics on orc turbine performance. J. Eng. Gas Turbines Power **135** (2013). 102603
6. Wheeler, A., Ong, J.: A study of the three-dimensional unsteady real-gas flows within a transonic orc turbine. In: ASME Turbo Expo, Dusseldrof, Germany (2014)

A Novel Acoustic Resonator for Speed of Sound Measurement in Dense Organic Vapours

Bertrand Mercier[✉], Nitish B. Chandrasekaran, and Piero Colonna

Propulsion & Power, Aerospace Engineering, Delft University of Technology, Delft, The Netherlands
{b.h.mercier,n.b.chandrasekaran,p.colonna}@tudelft.nl

Abstract. Speed of sound measurements are currently of the utmost importance for the development of thermodynamic models of fluids. A need for accurate thermodynamic models was identified for siloxanes, because of their use as working fluid in organic Rankine cycle power systems and because of the scientific interest in non ideal and non classical gas dynamics. The initiative of designing and realizing a novel acoustic resonator capable of performing speed of sound measurements in dense vapors at temperatures up to 400 °C stemmed from these considerations. Preliminary and unique experimental results are presented for siloxane D_6 (dodecamethylcyclohexasiloxane) for two isotherms at 346 °C and 348 °C and at pressures ranging form 5.5 bar to 6 bar. These measurements demonstrate for the first time the possibility of measuring the speed of sound of dense organic vapor at very high temperature. An initial comparison with speed of sound estimations provided by current thermodynamic models shows that the calculated values are approximately 6% higher.

Keywords: Siloxane · D_6 · Speed of sound · Acoustic resonator

1 Introduction

The accurate estimation of thermodynamic properties of fluids is often an essential requirement for engineering developments. The properties of most commonly used working fluids are currently formulated in terms of the Helmholtz energy if a sufficient number of reliable experimental data is available, and an outstanding precision can be achieved. Speed of sound measurements are particularly important because they can be highly accurate and are reach of thermodynamic information, as they can be used to correlate volumetric and caloric properties [8].

Innovation in organic Rankine cycle power systems technology largely depends also on the possibility of adopting novel working fluids, which better suit the various technological and non-technological requirements. Sufficiently accurate thermodynamic models are mandatory and their development depends

M. Pini et al. (Eds.): NICFD 2020, ERCOFTAC Series 28, pp. 162–168, 2021.
https://doi.org/10.1007/978-3-030-69306-0_17

on the availability of accurate measurements [1]. Furthermore, in recent years, significant steps forward were made in the niche field of non ideal compressible fluid dynamics (NICFD), all resting on the possibility of relying on accurate fluid thermodynamic models.

These applications demand for the estimation of properties at high reduced temperature and pressure in the dense vapor thermodynamic region, and experimental data in these conditions are not available, especially for complex organic molecules, given the challenges of performing measurements at high temperature.

The idea of conceiving and realizing a high-temperature resonator to measure the speed of sound of organic vapors in the dense vapor phase is motivated by these needs.

A comprehensive review of methods and rigs for speed of sound measurements commonly used to obtain data for the development of thermodynamic property models of fluids can be found in Ref. [6]. The speed of sound can be obtained either by measuring the travel time of a pressure perturbation along a known distance, or by detecting the induced resonance which builds up in cavities for a given acoustic wave number. Errors in the order of 0.01% can be achieved with both techniques [4,7,10].

Here, a speed of sound measurement apparatus designed to be operated at temperatures as high as 400 °C and pressure up to 10 bar is presented for the first time, together with its initial commissioning. Special attention is given to the design of the acoustic actuation system, which is particularly challenging at high temperature. A brief description of the operating procedures is also provided. Preliminary results related to the speed of sound of D_6 in a thermodynamic region exhibiting strong dense gas effects are also treated.

2 The Organic Vapor Acoustic Resonator (OVAR)

All types of acoustic resonators consist of a cavity filled with the fluid of interest and are fitted with at least two systems for generating and measuring acoustic waves. Here, the cavity serves the purpose of promoting a modal behaviour of the wave propagation, which is then used to determine the speed of sound from the resonant frequencies associated with the modes. For instance, considering the OVAR, which is a rectangular cuboid of length l_0 and square section of length h_0, the resonant frequencies are given as a function of the speed of sound c by

$$f_{k,m,n} = \frac{c}{2}\sqrt{\left(\frac{k}{l_0}\right)^2 + \left(\frac{m}{h_0}\right)^2 + \left(\frac{n}{h_0}\right)^2} \tag{1}$$

where k, m and n are integers corresponding respectively to the mode numbers in the longitudinal and the two radial directions. If l_0 is significantly larger than h_0, this relation can be simplified as

$$f_k = k\frac{c}{2l_0} \tag{2}$$

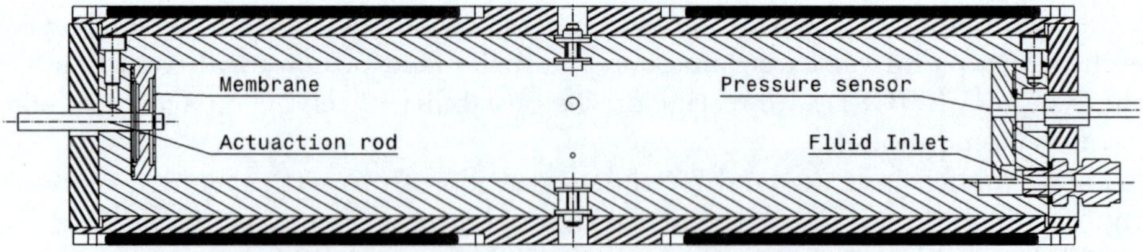

Fig. 1. CAD drawing of the longitudinal section of the resonator.

for values of k such that the frequency of the longitudinal mode is significantly smaller than that of the first radial mode. The dimensions of OVAR are $l_0 =$ 283 mm and $h_0 = 40$ mm, which allows to use Eq. (1) for $k \leq 6$. The frequency $f_{7,0,0}$ is indeed only 1.2% lower than $f_{0,1,0}$, which means that both modes may overlap.

Figure 1 shows a section view of the resonator. The cavity is machined from a block of stainless steel 316L, and closed on the top by a plate, also of stainless steel. The acoustic waves are generated by a 30 mm diameter membrane (labelled in the drawing), which is driven by a piezoelectric stack (Thorlabs PK3JUAP1) by means of a rod, at a frequency up 600 Hz. A high frequency pressure sensor capable of sustaining temperatures as high as 500 °C (Kulite XTEH-10L-190SM-300PSI-A) is fitted on the opposite side of the cavity. The fluid is injected through a small circular duct, labelled as fluid inlet in Fig. 1, with the help of a volumetric hand pump. The inlet pipe is also equipped with a pressure sensor (Druk Unik 5000) which is located in an environment at ambient temperature.

Heating is provided by four electric pads, in black in Fig. 1, delivering 50 W each and mounted on a 10 mm thick aluminium shell ensuring a sufficient temperature homogeneity all along the resonator. The temperature of each pad is measured and controlled within ±0.1 °C up to 400 °C with the help of four thermocouples inserted in the aluminium shell near the center of the corresponding pad. In addition, the fluid temperature temperature is measured by a K-type thermocouple of diameter 0.7 mm that protrudes into the resonator together with a NI-9210 thermometer. Both devices were calibrated for temperatures ranging between 175 °C and 375 °C. The thermal insulation is guaranteed by a 45 mm thick casing of ceramic wool (not shown in Fig. 1).

3 Application

3.1 Experimental Procedure

The resonator is operated by setting the thermodynamic state of the fluid along isotherms. The first step consists in heating the apparatus up to the set point, while vacuum is maintained in the cavity. The fluid is then injected using the hand pump. Once all the fluid is vaporised inside the cavity, a sequence is run to determine the frequency response of the resonator in order to calculated the

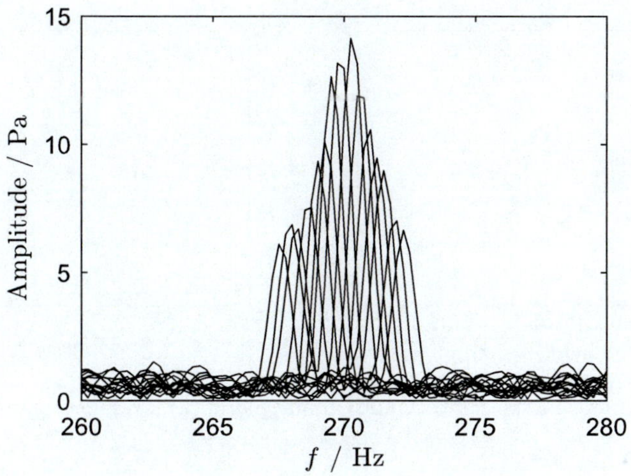

Fig. 2. Mode $k = 2$ acoustic resonance.

speed of sound by means of Eq. (2). The membrane is excited with a sinusoidal motion at a given frequency, while the acoustic pressure at the opposite end of the cavity is simultaneously recorded by the high frequency pressure sensor. This operation is repeated by shifting the excitation frequency across the frequency range that contains the resonance frequency of either mode $k = 2$ or $k = 3$. After a sequence, the fluid density is changed by injecting or remove some fluid.

For all measurement points, the fluid temperature inside the cavity is measured together with the pressure inside the inlet pipe. The density is determined by estimating the amount of fluid injected from the hand pump into the cavity, given that the volume V occupied by the fluid is known. Therefore, for any measurement point, the temperature, density, pressure and speed of sound are available.

3.2 Post Processing Procedure

The resonance frequency of the cavity is determined by post processing the series of acoustic signals recorded for all individual test points with varying excitation frequencies. The Fourier transform of each signal is computed to determine the amplitude of the acoustic signal at the prescribed frequency. Figure 2 depicts the pressure amplitude vs frequency recorded for an exemplary measurement point. In this example, the fluid was D_6 at a temperature T = 346 °C and a pressure of 5 bar. The excitation frequency was varied from 267 Hz to 272 Hz with steps of 0.33 Hz. A well defined peak is observed in the frequency response of the resonator, which indicates that a resonance has been captured. The resonance frequency f_0 is determined by fitting the top part of the peak with a Lorentzian function. The speed of sound is then deduced from this frequency using Eq. (2),

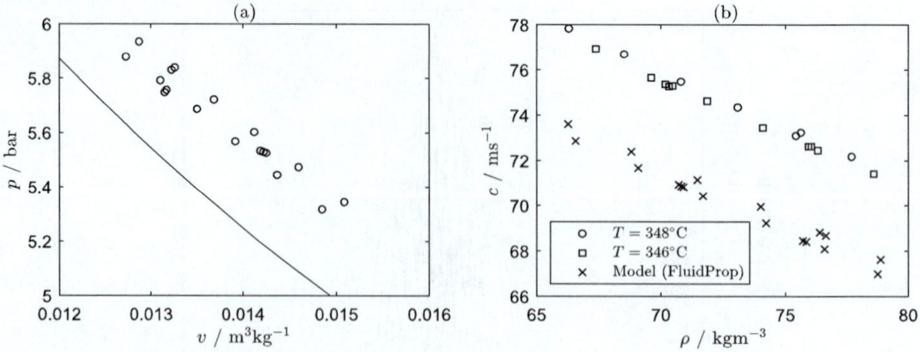

Fig. 3. Measurement in the Pv thermodynamic plane (a), and corresponding speed of sound (b).

Table 1. Measurements uncertainty.

Property	Uncertainty	Device
Temperature	± 0.5 K	k-type Thermocouple + NI–9210
Pressure	± 4 mbar + 0.1%	Druk UNIK 5000–16 bars
Density	$\pm 0.44\,\mathrm{kg/m^3}$	Custom hand pump
Speed of sound	$\pm 0.3\%$	Custom resonator

as

$$c = \frac{2lf_0}{k}, \tag{3}$$

where l is the length of the cavity corrected for the thermal expansion using $l = l_0(1+\alpha T)$, $\alpha \simeq 17^{-6}$ K^{-1} being the thermal expansion coefficient of stainless steel 316L.

3.3 Uncertainty Analysis

Speed of sound values are measured with acoustic resonators with the aim of assessing and improving thermodynamic models. It is thus of major importance to estimate the uncertainties associated with the state variables and with the speed of sound. Currently, no calibration of pressure and temperature sensors have been performed yet, therefore pressure and temperature uncertainty are estimated from the specs of the sensors and related data acquisition system: values are summarised in Table 1. The density is set by means of a hand pump of volume displacement $v_d = 1.060$ cm^3 per turn with a resolution of the order of 0.1 turn. This volume is converted into the value of the injected mass under the assumption that the density is accurately known. The density of the liquid in the pump is estimated as the density at 20 °C corrected for the temperature with the derivatives measured for D_5 [9]. At ambient conditions, the error introduced by the resolution is 0.1 g for a liquid density of $\rho = 960$ kg/m^3. In addition, the precise location of the transition from liquid to vapour in the inlet pipe is not known. This adds another 0.1 g of uncertainty to the estimation. After dividing these masses by the volume of the resonator $V = 455$ cm^3, the overall uncertainty for the density is ± 0.44 kg/m^3.

Typical errors affecting speed of sound measurements obtained with a resonator are extensively studied in the literature [3,5]. Among those biases, the most relevant for the OVAR setup arise from neglecting the influence of the fluid viscosity and should be smaller than 0.1%. Another type of error may be attributed to the deviation of the cavity shape form an ideal rectangular box. A finite element analysis of the modal behaviour of the cavity was carried out to provide further insights in this respect. The maximum deviation due to this error affecting the estimation derived from Eq. (3) was found equal to 0.29% for mode $k = 2$. This is a systematic error, which is compensated in the post processing of the measurement data. However, it is safe to use this figure as the speed of sound uncertainty to account for possible other shape imperfections.

3.4 Results

The first preliminary experimental campaign involved measurements of D_6 speeds of sound along the two isotherms at 346 °C and 348 °C, at densities ranging between $\rho = 66$ kg/m^3 and $\rho = 79$ kg/m^3. Figure 3(a) reports measurement points the Pv diagram of D_6. The thermodynamic states of these measurements were chosen such that they are close to the dew line and in a thermodynamic region exhibiting strong dense gas effects, as testified by the local value of the compressibility factor $Z \simeq 0.67$. Figure 3(b) shows the comparison between speed of sound measurements values computed with the multiparameter equation of state model documented in Ref. [2]. The comparison shows a very good agreement in terms of trends, however the model seems to underestimate the speed of sound by approximately 6%. This disagreement is much larger than the values of uncertainties reported in Table 1, therefore the possibility of using the OVAR to refine thermodynamic models of organic fluids is fully demonstrated by this first measurement campaign.

4 Conclusion

A novel resonator for high temperature speed of sound measurements in dense organic vapors was conceived, designed, realized and successfully tested in laboratories of the Propulsion and Power group at Delft University of Technology. Preliminary speed of sound measurements were performed in siloxane D_6 vapour in highly non ideal conditions with an estimated accuracy of $\pm 0.3\%$. This set of measurements provides a unique database which is compared with the predictions of what is arguably the best thermodynamic model available for this fluid. Significant discrepancies are observed for the thermodynamic conditions of interest. Further speed of sound measurements will be performed and improvement to the instrument and method implemented in order to provide a sufficient number of fully characterized experimental data aiming at the improvement of thermodynamic models of D_6 and possibly of the accuracy of the estimation of its fundamental derivative of gas dynamics in the region in which it is negative or close to zero. It is foreseeable that the OVAR will also be used with other fluids of interest in organic Rankine cycle technology research and non ideal compressible fluid dynamics.

Acknowledgment. The authors thank the Netherlands Organization for Scientific Research (NWO) which funded this research through the grant with project number 15837.

References

1. Colonna, P., et al.: Organic Rankine cycle power systems: from the concept to current technology, applications, and an outlook to the future. J. Eng. Gas Turb. Power **137**(10), 100801–19 (2015)
2. Colonna, P., Nannan, N.R., Guardone, A.: Multiparameter equations of state for siloxanes: $[(CH_3)_3\text{-Si-}O_{1/2}]_2\text{-}[\text{O-Si-}(CH_3)_2]_{i=1...3}$, and $[\text{O-Si-}(CH_3)_2]_6$. Fluid Phase Equilib. **263**(2), 115–130 (2008)
3. Gillis, K.A.: Thermodynamic properties of two gaseous halogenated ethers from speed-of-sound measurements: difluoromethoxy-difluoromethane and 2-difluoromethoxy-1,1,1-trifluoroethane. Int. J. Thermophys. **15**(5), 821–847 (1994)
4. Gillis, K.A., Moldover, M.R., Goodwin, A.R.H.: Accurate acoustic measurements in gases under difficult conditions. Rev. Sci. Instrum. **62**(9), 2213–2217 (1991)
5. Liu, Q., Feng, X., An, B., Duan, Y.: Speed of sound measurements using a cylindrical resonator for gaseous carbon dioxide and propene. J. Chem. Eng. Data **59**(9), 2788–2798 (2014)
6. Meier, K.: The pulse-echo method for high precision measurements of the speed of sound in fluids. Postdoctoral Thesis, Helmut-Schmidt-Universitat/Universitat der Bundeswehr Hamburg (2006)
7. Meier, K., Kabelac, S.: Speed of sound instrument for fluids with pressures up to 100 MPA. Rev. Sci. Instrum. **77**(12), 123903 (2006)
8. Span, R.: Multiparameter equations of state: an accurate source of thermodynamic property data. Springer Science and Business Media (2013). https://doi.org/10.1007/978-3-662-04092-8
9. Thol, M., Javed, M.A., Baumhögger, E., Span, R., Vrabec, J.: Thermodynamic properties of dodecamethylpentasiloxane, tetradecamethylhexasiloxane, and decamethylcyclopentasiloxane. Ind. Eng. Chem. Res. **58**(22), 9617–9635 (2019)
10. Zhang, J., Lin, H., Sun, J., Feng, X., Gillis, K., Moldover, M.: Cylindrical acoustic resonator for the re-determination of the Boltzmann constant. Int. J. Thermophys. **31**(7), 1273–1293 (2010)

First Experiments and Commissioning of the ORCHID Nozzle Test Section

Fabio Beltrame[1], Adam J. Head[1(✉)], Carlo De Servi[2], Matteo Pini[1], Ferdinand Schrijer[1], and Piero Colonna[1]

[1] TU Delft, Delft, Netherlands
{f.beltrame,a.j.head,m.pini,f.f.j.Schrijer,p.colonna}@tudelft.nl
[2] VITO, Mol, Belgium
carlo.deservi@vito.be

Abstract. This paper reports one of the initial NICFD experiments in the nozzle test section of the ORCHID aimed at providing accurate data for the validation of flow solvers, albeit, at this stage of the research, the focus is limited to inviscid phenomena. Notably, a series of schlieren photographs displaying Mach waves in the supersonic flow of the dense vapor of siloxane MM were obtained and are documented here for the commissioning experiment, namely, for inlet conditions corresponding to a stagnation temperature and pressure of $T_0 = 252\ ^\circ$C and $P_0 = 18.4$ bara. At these inlet conditions the compressibility factor of the fluid is $Z_0 = 0.58$. The digital processing of the schlieren images allowed to estimate multiple angles formed by the Mach waves stemming from the upper and lower nozzle surfaces because of the infinitesimal density perturbations generated by the, albeit small, roughness of the metal surfaces. These values are related to the local Mach number by a simple geometric relation. Moreover, the total expanded uncertainty in the Mach number was computed. This information together with the estimate of the average Mach number was used for a first assessment of the capability of evaluating NICFD effects occurring in a dense organic vapor flow of MM by comparison with the results of CFD simulations. The outcome of the comparison was satisfactory. It can thus be inferred that the nozzle test section has been commissioned and it is ready for experimental campaigns in which its full potential in terms of measurements accuracy, repeatability, and operational flexibility will be exploited.

Keywords: Schlieren measurements · Data processing · Error identification · Uncertainty estimation

1 Introduction

The computer simulation of compressible flows is critical for a large variety of scientific studies and engineering purposes. CFD codes have reached a remarkable level of accuracy. A vast amount of accurate experimental data is available for their verification and validation [10], as well as, in principle, to improve their accuracy via, e.g., calibration. However, all this body of experimental work deals to an almost complete extent with aerodynamics or classical and ideal gas dynamics. There is limited data available

M. Pini et al. (Eds.): NICFD 2020, ERCOFTAC Series 28, pp. 169–178, 2021.
https://doi.org/10.1007/978-3-030-69306-0_18

for geometries and flows typically encountered in ORC expanders and, more in general, for those constituting the research subject of Non-Ideal Compressible Fluid Dynamics (NICFD), such as transcritical expansions close to the critical point and compressible two-phase flows. Although first experimental results on non-ideal compressible effects in supersonic flows of dense organic vapours have been recently published, see Ref. [11, 12, 14], the data in the literature remain insufficient in terms of quantity and quality for the validation of modern CFD software. For this reason, the Organic Rankine Cycle Hybrid Integrated Device (ORCHID) has been conceived and realized to provide both accurate experimental data for paradigmatic NICFD benchmark flows and, in a second phase of the research, measurements for turbomachinery operating partly or entirely in the NICFD regime. The ORCHID is a continuous operating-type vapor tunnel capable of realizing a high-temperature organic Rankine cycle (ORC) with different working fluids in numerous operating conditions [7]. One of the test sections, namely, the nozzle test section, has been realized to 1) generate paradigmatic supersonic flows, similarly to what is done for classical and ideal gas dynamics, but for flows in the NICFD regime, namely a simple flow in a convergent divergent either empty or with an obstacle inserted to generate shock waves, and 2) to accurately measure quantities for validation of CFD codes capable of simulating NICFD flows. This paper describes the initial experiments carried out during the commissioning of the ORCHID nozzle test section. In particular, datasets have been generated for: i) the total conditions at the inlet and outlet of a planar nozzle, and, ii), thanks to schlieren imaging, Mach lines over almost the entire length of the diverging part of the nozzle. These quantities are sensitive to NICFD effects because of the influence of both volumetric and caloric thermodynamic properties of the fluid. Furthermore, all error sources influencing the measurement of the Mach number have been identified and subsequently quantified, including the uncertainties arising from the post processing method of the schlieren images.

The paper is structured as follows. Section 2 presents a description of the primary components of the nozzle TS. Furthermore, the schlieren layout and measurement procedure together with the specifications of the commissioning experiment are defined. In Sect. 3 the schlieren post processing method and the uncertainty estimation of the experiment error sources is presented. Section 4 reports the average midline Mach number distribution in the nozzle kernel region together with the associated uncertainties. The results are also compared with CFD predictions. The conclusions and an outlook for future work are presented in Sect. 5.

2 Nozzle Test Section: Layout and Schlieren Measurement Procedure

Figure 1a shows an isometric view of the ORCHID nozzle test section (TS), which consists of three primary compartments: the settling chamber (SC) which contains the flow conditioning devices such as honeycombs and screens; the nozzle housing (NH), which hosts a planar converging-diverging nozzle; and, the receiver (RC). The nozzle is made from two equal and removable profiles, which are inserted at the top and bottom of the nozzle housing. Figure 1b shows the diverging portion of this profile which has been designed with the method of characteristics [5]. The side walls of the nozzle

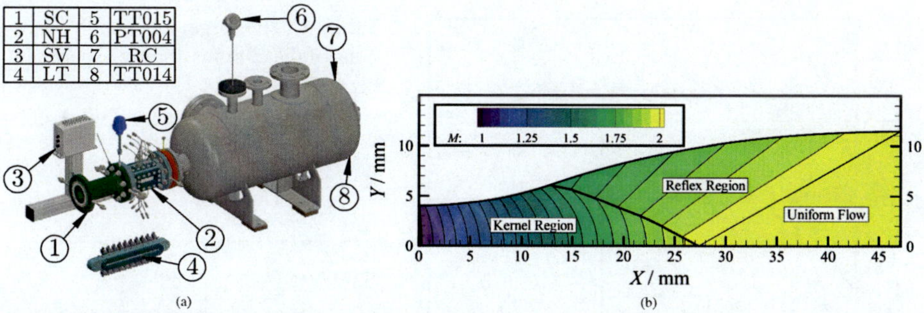

Fig. 1. (a). The primary components of the nozzle TS. SC: settling chamber; NH: nozzle housing; SV: Scanivalve; LT: liquid traps; RC: Receiver. **(b).** Flow Mach number in the diverging portion of the nozzle as calculated with the method of characteristics [5] for the MM working fluid at the design conditions: $T_0 = 252$ °C, $P_0 = 18.4$ bar and a back pressure of 2.1 bar. Taken from [6]. Fluid properties estimated with FLUIDPROP [4] Model = REFPROP [8].

are made from transparent windows which provide optical access. The test section is equipped with a Scanivalve (SV) DSA3218 pressure scanner which is used, in combination with pressure taps, to characterise the total pressure in the settling chamber and to measure the static pressures along the nozzle profile. To avoid the accumulation of condensate in the lines connecting the pressure taps to the Scanivalve, the pressure scanner is equipped with a purging system. Moreover, a specially engineered liquid trap (LT) allows to visualise vapor condensation during pressure measurements. As far as the total temperature is concerned, this is measured at the nozzle inlet (TT015) with a *Wika* PT100 transmitter. The expanded uncertainty is of Class A, i.e., $\pm 0.15 + 0.002$ $|T|$ °C as per the EN 60751 standard. The same type of transmitter is also used to measure the temperature (TT014) in the receiver. Finally, the back pressure (PT004) is measured by a *Wika* UPT 20 pressure transmitter with an expanded uncertainty of ± 0.1 bar.

The adopted schlieren layout is a two-lens flat mirror z-type configuration. The schlieren equipment, then, consists of a light source (cold white LED from Thorlabs), a pinhole (diameter 0.2 mm), two collimating and schlieren lenses (diameter 116 mm, focal length 1524.7 mm), two mirrors $(L \times H = 127 \times 178)$ mm, a knife edge and a camera. This equipment is mounted on three separate rigid structures which ensure the correct alignment of all components, see Ref. [6]. The camera used to image the flow field is a BOBCAT IGV-B1610 16bit CCD camera with a maximum resolution of 1628×1236 pixels.

The operating conditions chosen for the commissioning experiments corresponds to those considered as design point of the ORCHID Balance of Plant (BoP) and nozzle test section. The targeted conditions are reported in Table 1 while Fig. 2 displays the thermodynamic states identifying the expansion process in the nozzle under the hypothesis of an isentropic adapted flow. When the ORCHID BoP warm-up is completed and the operating conditions have stabilized around the values of total inlet pressure and temperature chosen for the experiment, the nozzle test section is opened. Around twenty

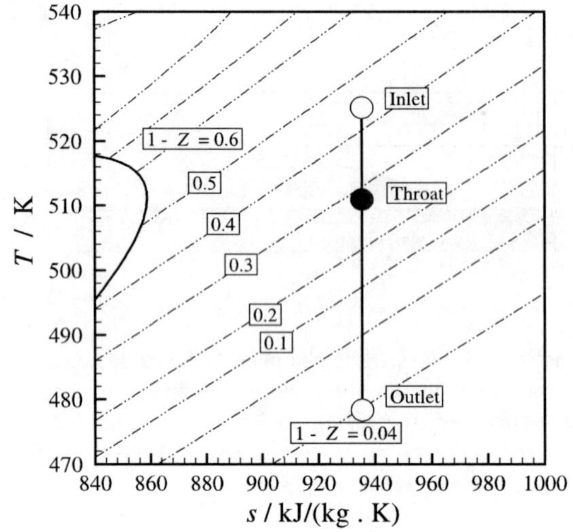

Fig. 2. $T - s$ diagram showing the isentropic expansion of the commissioning experiment with contours of 1-Z.

Table 1. The target boundary conditions of the commissioning experiment PR.025-NT.002.

Fluid	Tot. temp. TT015/°C	Tot. press. P_{SV}001/bara	Comp. F. Z_t
MM	252	18.4	0.56
Back p. PT004/bara	Outlet Mach. M_2	Mass flow FT001/kg/s	
2.1	2	1.145	

minutes are then required to stabilize again the nozzle inlet conditions, provided that the test section was preheated before to sufficiently high temperature (say 180 °C for NICFD experiments with MM). During this settling period, the back pressure PT004 of the nozzle is set to the desired value. Afterwards, the schlieren image acquisition is activated. An image acquisition frequency 25 Hz is appropriate and allows to achieve statistical convergence of the measure quantities, provided that the flow field and boundary conditions are sufficiently steady. Approximately 2,000 images are recorded in order to characterise the Mach number along the midplane of the nozzle. Then the inlet and outlet valves of the nozzle test section are closed and an image of the empty nozzle is taken for the postprocessing of the schlieren images.

3 Schlieren Images Postprocessing and Measurement Uncertainty

The method for the Mach number extraction and uncertainty evaluation from a schlieren image dataset consists of the following main steps: first, the entire image dataset is enhanced by aligning each single frame to the first schlieren image to compensate for possible shifts and rotations in the reference frame of each picture caused by vibrations in the test section. This operation is performed in the DaVis software 8.3.0 from Lavision. Second, a Mach line extraction algorithm implemented in Matlab is applied to evaluate the slope of the Mach waves with respect to the local flow velocity (Mach line angle μ) at different axial positions along the nozzle midplane for a sequence of schlieren images. The code evaluates also the total expanded uncertainty associated with the estimated Mach line angles. The accuracy of the extracted information is, then, assessed by verifying flow symmetry and steadiness. The flow is deemed symmetric if the difference in the angles of the Mach lines converging to the same location along the nozzle midplane is smaller than a given threshold, i.e. the smaller total expanded uncertainty estimated by the tool for the angles of the two intersecting Mach lines. As far as the steadiness of the flow is concerned, this check is performed by verifying that the mean and standard deviations of the Mach line angles extracted by the algorithm remain constant as the number of processed images increases. For the process run under consideration, the number of images required to obtain a constant mean and standard deviation is approximately one thousand. Given the acquisition frequency of the camera, i.e. 25 Hz, this corresponds to the analysis of 40 s of flow history. Finally, if the flow symmetry and steadiness is verified, the estimated Mach line angles in the diverging part of the nozzle are used to calculate, for each image, the corresponding Mach number distribution according to the Mach equation

$$M = \frac{1}{\sin(\mu)}. \tag{1}$$

Describing more in detail the Mach line extraction algorithm, the first operation consists in applying a 2D Gaussian filter to the processed image in order to remove noise, like darker and lighter spots, which might be due, for example, to spurious reflections. Subsequently, the Canny edge detection algorithm is used to identify the contours of the Mach lines, by computing the intensity gradients of the image with respect to a given threshold. The outcome of the Canny algorithm is a binary image which is then divided into sub-images called interrogation windows centered on the nozzle midplane. The Mach lines tend to bend at increasing distance from the midplane due to 2D flow effects, whose intensity augments the closer the location to the nozzle throat. The interrogation window dimension is, then, adapted to ensure that the maximum line curvature is lower than the angular resolution, which implies that no curvature effect are detectable in the portion of the image processed by the algorithm. This is achieved by specifying a window height to nozzle height ratio equal to 0.15. In the experiment at hand, the size of the interrogation windows (rows $\times$ columns), then, ranges from 45×50 pixels in the throat region to 60×70 pixels at the nozzle exit. Finally, the Hough transform [9] is applied to identify the lines which best fit the contours or edges detected by means of the Canny algorithm. More in detail, the Hough transform determines for each pixel of the selected interrogation window if a straight line with a specific orientation and distance from the origin r may pass through that pixel. The result of this process is a two-dimensional matrix called accumulator which ranks the different lines based on the number of pixels they fit. The number of columns and rows of the matrix is a function of the resolution set by the user to discretize, respectively, the inclination θ and the distance from the origin r of the lines. The greater the resolution, the larger the matrix. The cell in the matrix with the highest value represents the combination of θ and r which best fits the edges determined by the Canny algorithm and that identifies a candidate Mach line. In order to ensure the detection of a Mach line and not a random number of collinear points, only the line with the highest value in the accumulator and with a length above a specified threshold is selected. For more details about the implementation of the Mach line angle detection algorithm, the interested reader is referred to Ref [2].

Figure 3a shows the average Mach lines superimposed on the first schlieren image. The domain of interest is the so-called kernel region of the supersonic flow, where the non-ideal gas dynamic effects are more prominent. Notice that the refractive index changes significantly due to the large variation in fluid density between the throat and the end of the kernel region, which translates to a substantial change in contrast. It occurs that the image tones become darker and darker moving from the nozzle outlet towards the inlet, up to the point that the light deviation exceeds the schlieren system measuring range in the area close to the throat. For this reason, it was not possible to estimate the flow Mach number in that region of the images. The contrast is insufficient to allow a consistent identification of the Mach lines over the dataset: the computed standard deviation of line angles in interrogation windows close to the throat exceeds a predefined threshold, namely three times the total expanded uncertainty calculated for the Mach line angles in the window that follows the one of interest with respect to the flow direction.

Primary error sources in the measurement of the flow Mach number include i) a misalignment of the optical system, ii) the finite and discrete resolution of the CCD sensor of the camera, and iii) approximations introduced by the implemented post-processing algorithms. In this regard, the total expanded uncertainty in the flow Mach number can be expressed as

$$U_M = \sqrt{(U_{\mathrm{M,data}})^2 + (U_{\mathrm{M,ext}})^2 + 2\rho_{\mathrm{data,ext}} \cdot U_{\mathrm{M,data}} \cdot U_{\mathrm{M,ext}}}\,, \tag{2}$$

where $U_{\mathrm{M,data}}$ is the uncertainty due to random effects, while $U_{\mathrm{M,ext}}$ is the uncertainty related to the algorithm used for processing the schlieren images and $\rho_{\mathrm{data,ext}}$ is the Pearson correlation coefficient between the two error sources. $U_{\mathrm{M,data}}$ can be estimated based on the standard deviation of the flow Mach angle resulting from the statistical processing of the whole image dataset and assuming a 95% confidence interval. $U_{\mathrm{M,ext}}$ can be evaluated as a function of the corresponding uncertainty in the measurement of the Mach line angle. In particular, $U_{\mathrm{M,ext}}$ may be calculated by propagating the uncertainty $U_{\mu,\mathrm{ext}}$ through the Mach equation, i.e.

$$U_{M,\mathrm{ext}} = M\sqrt{M^2-1}U_{\mu,\mathrm{ext}}, \tag{3}$$

where the multiplicative factor for $U_{\mu,\mathrm{ext}}$ is derived by evaluating the derivative of the Mach number with respect to the Mach line angle $|\frac{dM(\mu)}{d\,\mu}|$ [2]. Equation 3 implies that the uncertainty on the flow Mach number increases from the throat to the exit of the nozzle. The uncertainty associated with the estimation of the Mach angle $U_{\mu,ext}$ is, in turn, estimated as

$$U_{\mu,ext} = \sqrt{AR_{95}^2 + U_{\mathrm{Hough,res}}^2}, \tag{4}$$

where $U_{\mathrm{Hough,res}}$ is the Hough transform (see [9]) angular discretization uncertainty, which is assumed to be half of the discretization step for the line orientation $\Delta\theta$ considered in the accumulator. On the other hand, AR_{95} is the 95% confidence level uncertainty on the detected angle due to the image resolution, or, in other words, the angular resolution associated with a given interrogation window. The finite number of pixels of the image inherently introduces an approximation on the line angle, which tends to increase the shorter the Mach lines [2]. As the size of the interrogation windows is not constant, the length of the detected Mach lines varies, thus the minimum angle that can be identified. The angular resolution is, indeed, equal to the arctangent of the inverse of the detected line length. AR_{95} is, then, determined for each interrogation window by calculating the average angular resolution and adding to this term two times the line length standard deviation propagated through the angular resolution equation.

Also the position at which the flow Mach number is measured is affected by uncertainty. First, distances in the image are in pixels and need to be converted into linear distances. The uncertainty associated to this operation depends on the adopted pixel-to-meter conversion ratio, here evaluated by dividing the throat height, measured with the caliper before the experiment, by the length in pixels of the channel height estimated in correspondence of the throat. Second, a Mach line identified by the Hough transform is *a priori* assumed passing through the center of the interrogation window processed by the algorithm. The corresponding uncertainty, assuming a 100% containment level and a uniform distribution, is equal to half of the interrogation window width.

4 Experiment Versus Numerical Simulation

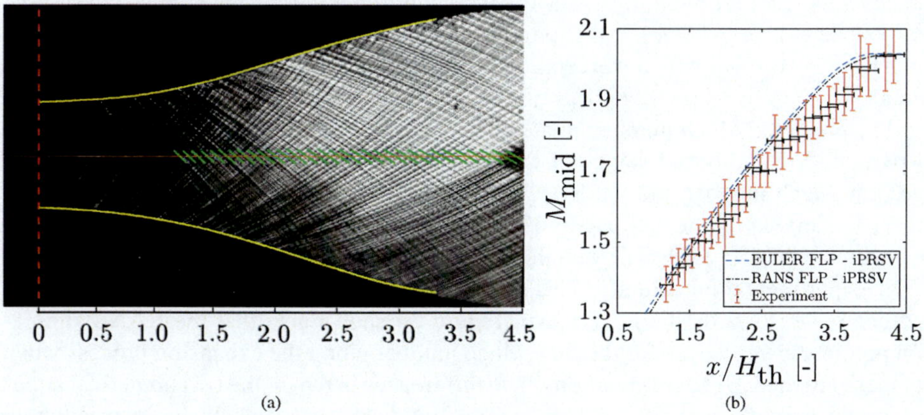

Fig. 3. (a). Average extracted Mach lines superimposed on the first schlieren image. **(b).** Comparison between the measured flow Mach number with the corresponding total expanded uncertainty and two different numerical predictions. Taken from [6].

Figure 3b shows the comparison between two numerical predictions against the Mach number distribution along the nozzle mid-plane obtained experimentally. The experimental data is complemented with the estimated total expanded uncertainty. The axial coordinate x in the figure, which represents the distance along the nozzle mid-plane from the throat, is normalized with respect to the nozzle throat height H_{th}. The total expanded uncertainty on the extracted Mach number ranges from $M = 1.38 \pm 0.05$ to 2.03 ± 0.082 at positions $x/H_{\text{t}} = 1.18$ and 4.33, respectively. The numerical solutions of the flow-field in the ORCHID nozzle are obtained using the open-source flow solver

SU2 for a 2D half-nozzle with symmetry conditions along the midplane. The generalized Roe scheme with second-order MUSCL reconstruction is used for the computation of the advective fluxes. The total inlet conditions assumed in the CFD model correspond to the average total pressure and temperature measured at the nozzle inlet during the experimental run once steady flow conditions established. Their values are $P_{0,\mathrm{avg}} = 18.36 \pm 0.18$ bar, $T_{0,\mathrm{avg}} = 252.4 \pm 0.69$ °C. The pressure measured at the nozzle outlet is $P_{\mathrm{b,avg}} = 2.06 \pm 0.25$ bar. The thermo-physical properties of the siloxane MM have been computed by an external fluid properties library [4]. In particular, the thermodynamic properties of the fluid are estimated by using the iPRSV cubic equation of state. The transport properties are calculated with the Chung's transport model. Furthermore, a fourth order polynomial function of the temperature is used for the ideal gas isobaric heat capacity [13]. For the inviscid simulations, the Euler implicit time marching scheme with a CFL of 20.0 is used. A grid convergence study based on the methods given in Ref. [1,3] was performed to determine a proper size of the unstructured mesh. This indicates that a grid of 10,000 elements provides mesh independent results and, as such, the numerical uncertainty can be deemed negligible. Regarding the RANS simulation, turbulence effects have been modeled by the Shear-Stress Transport (SST) closure. The calculations have been run with a high-order resolution scheme for the advective and turbulent fluxes. Moreover, additional mesh cells are used to resolve the boundary layer with $y^+ < 1$ along the nozzle walls. Finally, all the CFD simulations were performed with a workstation equipped with an Intel i7-8750H (2.21 GHz, 6 cores).

The predicted Mach numbers match reasonably well with the experimental results, in particular at the initial stages of the expansion. The difference between the average flow Mach number and the RANS prediction at position $x/H_{\mathrm{th}} = 1.2$ is less than 0.6%. The maximum deviation is 4.4% of the experimental Mach number, and occurs at position $x/H_{\mathrm{th}} = 3.54$. The simulation results are slightly above the upper bound of the experimental uncertainty. This deviation tends to increase as the Mach number augments, i.e. towards the nozzle exit. Figure 3b shows also that the RANS simulation predicts lower values for the flow Mach number along the expansion process when compared to the EULER simulations. The differences between the two numerical simulations are, however, very small. The reasons for the discrepancy between the prediction of the CFD models and the experimental data are most likely owed to the fact that, the former do not account for 3D effects, such as the boundary layer development on the flat glass window panes, which causes a reduction in the nozzle area ratio. Furthermore, thermal effects are not negligible. A different dilation of the nozzle housing with respect the nozzle profile may alter the nozzle throat height during the test runs, thus modifying the nozzle flow area. This effect will be assessed in future experiments.

As it can be noticed from Fig. 3b, the total uncertainty on the flow Mach number increases towards the exit of the nozzle. This occurs despite the random uncertainty $U_{\mathrm{M,data}}$ is almost constant in the analyzed portion of the flow, with a slight increase towards the nozzle throat where lines are less visible and the quality of the images decreases. Also the uncertainty associated to the Mach line angles, U_{μ}, is almost constant, at least as long as the brightness level of the image is constant and the images are sharp. The reason of the trend in the total uncertainty in Fig. 3b is, then, the increase of

$U_{M,ext}$, in accordance with Eq. 3, which shows that uncertainties in the μ angle increase with the flow for Mach number for values of this quantity greater than 1.3.

5 Conclusions

This paper reported initial results of the ORCHID nozzle commissioning experiments. A sequence of schlieren images allowed to characterise the Mach lines associated with the expansion of MM in a thermodynamic region where the properties of the vapour substantially deviate from those of an ideal gas. An *ad-hoc* methodology has been implemented to process the schlieren images and to estimate the flow Mach number on the nozzle midplane and its total uncertainty. These data have been, then, used to make a preliminary assessment of the predictive capabilities of the SU2 flow solver for NICFD applications. In particular, predicted Mach numbers matched well with the experimental results at the initial stages of the expansion. Deviations towards the end of the kernel region may be due to viscous phenomena not accounted for in the CFD models and thermal dilation effects in the nozzle test section which may alter the throat area. It was also observed that the uncertainty on the flow Mach number evaluated experimentally strongly depends on the uncertainty associated with the post-processing of the schlieren images, where the main source of error is the limitation on the angular resolution of the Mach lines which, in turn, depends on the pixel resolution of the images and the size of the Hough transform search space.

Acknowledgments. The authors would like to express their gratitude to Liam Bills for producing the simulation results with SU2.

References

1. American Society of Mechanical Engineers (ASME): Standard for verification and validation in computational fluid dynamics and heat transfer. Technical report V&V20-2009, ASME (2009)
2. Beltrame, F.: Accuracy assessment of the SU2 flow solver for non-ideal organic vapor supersonic expansions using experimental data. Master's thesis, TU Delft, Politecnico di Torino (2020)
3. Celik, I.B., Ghia, U., Roache, P.J., Freitas, C.J., Coleman, H., Raad, P.E.: Procedure for estimation and reporting of uncertainty due to discretization in CFD applications. J. Fluids Eng. **130**(7), 7 (2008)
4. Colonna, P., van der Stelt, T.P., Guardone, A.: FluidProp (version 3.0): A program for the estimation of thermophysical properties of fluids (2012)
5. Guardone, A., Spinelli, A., Dossena, V.: Influence of molecular complexity on nozzle design for an organic vapor wind tunnel. J. Eng. Gas Turbine Power **135**(4), 042307 (2013)
6. Head, A.J.: Novel experiments for the investigation of non-ideal compressible fluid dynamics: the ORCHID and first results of optical measurements. Ph.D. thesis, Delft University of Technology (2021)
7. Head, A.J., De Servi, C., Casati, E., Pini, M., Colonna, P.: Preliminary design of the ORCHID: a facility for studying non-ideal compressible fluid dynamics and testing ORC expanders. In: ASME Turbo Expo, number GT2016-56103, p. 14 (2016)

8. Lemmon, E.W., Huber, M.L., McLinden, M.O.: NIST standard reference database 23: reference fluid thermodynamic and transport properties-REFPROP, version 9.1. Technical report, National Institute of Standards and Technology, Standard Reference Data Program, Gaithersburg (2013)
9. Lo, R., Tsai, W.: Gray-scale Hough transform for thick line detection in gray-scale images. Pattern Recogn. **28**(5), 647–661 (1995)
10. Roache, P.J.: Verification and Validation in Computational Science and Engineering. Hermosa Publishers, Albuquerque (1998)
11. Robertson, M., Newton, P., Chen, T., Costall, A., Martinez-Botas, R.: Experimental and numerical study of supersonic non-ideal flows for organic rankine cycle applications. J. Eng. Gas Turbines Power **142**(8) (2020). https://doi.org/10.1115/1.4046758
12. Spinelli, A., Cammi, G., Gallarini, S., Zocca, M., Cozzi, F., Gaetani, P., Dossena, V., Guardone, A.: Experimental evidence of non-ideal compressible effects in expanding flow of a high molecular complexity vapor. Exp. Fluids **59**(8), 126 (2018). ISSN 1432-1114
13. van der Stelt, T.P., Nannan, N.R., Colonna, P.: The iPRSV equation of state. Fluid Phase Equilibria **330**, 24–35 (2012)
14. Zocca, M., Guardone, A., Cammi, G., Cozzi, F., Spinelli, A.: Experimental observation of oblique shock waves in steady non-ideal flows. Exp. Fluids **60**(6), 101 (2019)

Author Index

M. Pini et al. (Eds.): NICFD 2020, ERCOFTAC Series 28, pp. 179–180, 2021.
https://doi.org/10.1007/978-3-030-69306-0